Industrializing America

THE AMERICAN MOMENT

Stanley I. Kutler
SERIES EDITOR

The Twentieth-Century American City, second edition
Jon C. Teaford

American Workers, American Unions, 1920–1985, 2d edition
Robert H. Zieger

A House Divided: Sectionalism and Civil War, 1848–1865
Richard H. Sewell

Liberty under Law: The Supreme Court in American Life
William M. Wiecek

Winning Is the Only Thing: Sports in America since 1945
Randy Roberts and James Olson

America's Half-Century: United States Foreign Policy in the Cold War,
2d edition
Thomas J. McCormick

American Anti-Communism: Combating the Enemy Within. 1830–1970
Michael J. Heale

The Culture of the Cold War
Stephen J. Whitfield

America's Welfare State: From Roosevelt to Reagan
Edward D. Berkowitz

The Debate over Vietnam
David W. Levy

And the Crooked Places Made Straight:
The Struggle for Social Change in the 1960s
David Chalmers

Medicine in America: A Short History
James H. Cassedy

The Republic of Mass Culture:
Journalism, Filmmaking, and Broadcasting in America since 1941
James L. Baughman

The Best War Ever: America and World War II
Michael C. C. Adams

Uneasy Partners: Big Business in American Politics, 1945–1990
Kim McQuaid

America's Right Turn: From Nixon to Bush
William C. Berman

Industrializing America: The Nineteenth Century
Walter Licht

INDUSTRIALIZING
AMERICA

THE NINETEENTH CENTURY

Walter Licht

THE JOHNS HOPKINS UNIVERSITY PRESS
Baltimore and London

© 1995 The Johns Hopkins University Press
All rights reserved. Published 1995
Printed in the United States of America on acid-free paper
04 03 02 01 00 99 98 97 96 95   5 4 3 2 1

The Johns Hopkins University Press
2715 North Charles Street
Baltimore, Maryland 21218-4319
The Johns Hopkins Press Ltd., London

Library of Congress Cataloging-in-Publication Data will be
found at the end of this book.
A catalog record for this book is available from the
British Library.

ISBN 0-8018-5013-4
ISBN 0-8018-5014-2 (pbk.)

To Lois and Emily

Contents

Editor's Foreword

THROUGHOUT THE WESTERN WORLD in the nineteenth century, progress and material change amounted to a form of secular worship. Certainly, the United States that Thomas Jefferson led in 1800 was a far different place in every respect from the one that Theodore Roosevelt served as President just after the twentieth century opened. Consider a continental-wide landscape, dominated by railroads, large-scale manufacturing, large commercial farms, and thriving urban centers; marked by lively trade and enterprise, as opposed to an earlier one largely confined to the eastern seaboard; and dotted by rough, expensive roads, specialized, small-scale manufacturing, subsistence farming, and small, isolated towns largely servicing an agrarian economy. In 1800, the United States was a new, vastly underdeveloped nation on the fringe of the western world; a century later, it stood at the threshold of commercial and financial supremacy, and it was an emerging military power.

The economic transformation in turn led to profound social and cultural changes. What caused this transformation, and what its implications were, is the subject of Walter Licht's exciting new synthesis and interpretation of that dramatic period. Licht takes us beyond the inclusive, yet always elusive, terms of "industrialization" and "urbanization." Instead, he applies an understanding to the nineteenth century of what we see as so commonplace in our own time, namely that all aspects of life, whether material things such as land and labor or notions of leisure, are subject to market considerations. "Everything," Licht writes, "became valued accordingly by the calculus of supply and demand and the cash nexus." "Everything" is no exaggeration. Demand for goods and services drove the market as Americans relentlessly pursued "profit," which itself could be either market or pleasurable. This notion of a market-oriented society hardly was unique or new to the nineteenth century. The very beginnings of American society in the seventeenth century involved similar considerations. But what changed was its pervasiveness and how it was or was not managed.

Finally, Licht confronts the meaning of all that change. He gauges the emerging commercialization of the early nineteenth century as a product of certain societal changes, but he also is acutely aware that such change often is the agent of further change. What is clear is that, as the nation entered the new twentieth century, it created a new political and economic order, one that partially remains with us but in some respects is totally unrecognizable as we mark the milestone of another century.

Stanley I. Kutler
THE UNIVERSITY OF WISCONSIN

Preface

●────────────●

I FACED several challenges in writing this book. A comprehensive text on industrialization in the United States in the nineteenth century has not been written. There are separate studies of the first and second halves of the century and the specialized works of economic, business, and labor historians and historians of technology. I assumed as my task the preparation of a century-long, cross-regional, synthetic book. Having no prior example to follow or revise provided a first challenge.

Comprehensiveness proved a second issue. The last twenty-five years have witnessed great upsurges in the writing of history. Methods and interpretations in old fields of study have been vitally transformed and new fields—such as urban and women's history—have appeared out of the proverbial thin air to achieve central prominence. I wanted to write a book that reflected and absorbed all the latest developments.

Interpretation represented a third matter. Vast disagreements exist among historians in all fields of study. I wanted to present to my readers the various ways in which scholars explain events and change. However, I did not want to encumber the text with lists of names of the various parties to disputes. Critical arguments appear in the book along with my particular judgments; readers in need of a scorecard can find one in the bibliographic essay that is provided at the end.

I also wanted this study to be informative and readable. The book would include information on traditional as well as new subjects of interest and could be used as an encyclopedia. Readers can learn in these pages about John D. Rockefeller as well as women shoemakers in Lynn, Massachusetts. I have tried as well to keep the book as jargon-free as possible.

Finally, there was the problem of the "glue," or what would hold the text together, an issue made all the more difficult without a prior model to rely on. The book does have a central theme—not a simple one—which is presented fully in the introduction that follows.

I was then operating somewhat in the dark in writing this book, but the above challenges made this an intriguing and exciting exercise. However, I was not totally without assistance. The book received fine readings from Michael Bernstein, Gary Kornblith, and Thomas Dublin, and I am indebted for their suggestions. I also received great encouragement from my editors at the Johns Hopkins University Press, Henry Tom and Stanley Kutler; they deserve awards for patience. Finally, I have dedicated this book to my wife and daughter—a small token of my love and appreciation.

Introduction

THE YEAR IS 1801. Imagine that foreign guests to the inauguration of Thomas Jefferson, third president of the United States, are handed a guidebook to the young republic. In reading, they learn that the new nation comprises 5.3 million people. The inhabitants of the country have taken title to the lands of an additional 600,000 native peoples—their numbers now greatly depleted from upwards of five million through two centuries of disease and warfare. The new claimants to the continent among themselves also form a relatively heterogeneous lot. Immigrants from England constitute 50 percent of the population and slaves from Africa, 20 percent; there are small but noteworthy communities as well of Germans, Scots, Irish, Dutch, French Huguenots, and Swedes. A few pioneering souls have settled west of a line of mountains that parallels the eastern coastline 150 miles from the shore, but the bulk of the population lives along the Atlantic seaboard.

Despite occupying a narrow band, the citizens of the new country reside in historically distinct regions. In New England small family farms abound, family members producing goods mainly for their direct consumption. A more commercialized and prosperous agriculture marks the Middle Atlantic states, with grain production on large homesteads. In the American South most visible are the huge tobacco, rice, and emerging cotton plantations worked by gangs of slaves and whose product is marketed throughout the Western world. There are also burgeoning seaport cities, such as New York and Philadelphia, where merchants orchestrate the outward and inward flows of agricultural and manufactured goods and artisans ply their trades. By the time of Thomas Jefferson's ascension to the highest office of the land, however, a mere 10 percent of the inhabitants of the young nation dwell in urban areas. This is a decidedly agrarian and rural society.

The year is now 1901. Imagine that guests to the second inauguration of William McKinley, twenty-fifth president of the United States, are presented with a new gazetteer. The physical boundaries of the country now encompass a vast domain. More than 77 million people live in forty-five states, which stretch from the Atlantic Ocean to the Pacific; during McKinley's first term in office the nation has even come into possession of foreign territories and is a rising world diplomatic and economic power. The Americans are now remarkably heterogeneous. Americans of German and Irish descent far outnumber those who identify with the English roots of the society; there are parts of the country in the Midwest where the population is distinctly Scandinavian; Chinese and Japanese communities dot the Pacific states, and Mexican communities, the Southwest; native Americans, now only 250,000 strong after further defeats and at the nadir in their population history, inhabit reservation lands largely in western states; and as President McKinley is delivering his second inaugural address, millions of newcomers with strange languages and customs from southern and eastern Europe are streaming into the cities of the East. Regional differentiation still marks the country, but there are now many more regions: there is logging in the Northwest, large-scale fruit and vegetable farming in California, mining in the mountain states, ranching in the Southwest, corn and wheat growing in the central prairies, dairying in the Upper Midwest, cotton still dominating in the South, prosperous mixed farming in the Ohio Valley, and industry in the Northeast.

A map in the gazetteer denotes these regions, and the unknowing viewer might mistakenly conclude that the nation remains primarily agrarian and rural; regions do tend to be defined and symbolized on maps by what the land bears. Agriculture indeed does remain critical. As the modern guidebook would indicate, more than 40 percent of the American people still work on farms in 1901, and this greatly affects the politics of the day. American farm goods amount to one-fourth of the country's gross national product, and they continue to be prime income producers in the world market economy. But that is not what the gazetteer would emphasize. The key story in 1901 is that of American industry, for as William McKinley accepts the accolades of the crowd in Washington, D.C., the United States stands as the world's leading industrial nation, manufacturer of one-third of the world's industrial output. One in four Americans now hold jobs in industry, and another 25 percent work in white-collar and service-sector jobs attendant to and spawned by industrial production. This new activity, of course, does not transpire in the countryside. Commerce bred cities, and now increased commerce from industry and industry itself spur sizable urban growth. At the turn of the twentieth century, 40 percent of the American people live in cities, and 20 percent in truly large cities with populations of more than 100,000 people. In 1901, the United States is a thriving industrial and urban nation.

This book is concerned with the great social and economic transformations that occurred in this country over the course of the nineteenth century between the ages of Jefferson and McKinley. When and where change occurred and the pace of change will be of prime importance, but the great issue will be the "why" of change. What caused America to be so fundamentally transformed?

Words may prove more of a hindrance than a help in our search for explanations, obscuring more than they enlighten. Scholars and non-scholars alike tend to invoke simple terms or phrases that seemingly explain why change occurred; usually these words just describe what happened rather than why it happened. Take the word used in the title of this book, *industrialization*. Why did the United States transform itself in the nineteenth century? A quick and common response would be, "Because the nation industrialized." Again, that possibly describes what happened, but the term has limited explanatory value. Why did the nation industrialize?

Even as a descriptive term, *industrialization* is faulty. *Industrialization* variously refers to the general shift from agriculture to manufacture, the rapid and widespread adoption of mechanical means of production and inanimate forms of energy, the spread of the wage labor system, and the coming of large factories. To say that the nation industrialized implies that these changes happened uniformly and evenly and constituted the key development of the times. The easy invocation of the term misses too much—that throughout the nineteenth century, for example, American farmers vastly expanded the agricultural base of the nation and that agriculture continued to represent an essential building block of the economy; that whole areas of the country remained relatively untouched by industrialization; that even in industrialized regions, the process unfolded sporadically and without pattern. Home production and craft and small and medium-sized shop work persisted alongside and were stimulated by the mammoth new factories and plants. Office and service employment emerged and expanded as well. The nation did industrialize, but the term does not fully capture the complexity or even the essence of change.

Urbanization is a similarly flawed term. Greater numbers of Americans came to reside in cities, but that is only a part of the story. *Commer-*

cialization is perhaps a more useful term. Why did the United States change so dramatically between the presidential administrations of Jefferson and McKinley? Because the nation "commercialized," or somewhat differently and less elegantly, became "capitalistic." The fundamental thought here is that America became an unfettered market society: that all production and consumption became totally oriented toward selling and buying in the marketplace and that everything—goods, land, labor, even time—became valued accordingly by the calculus of supply and demand and the cash nexus. Market activity drove change; increased market activity even caused industrialization (and urbanization, for that matter).

Pointing to the expansion of the market as the key engine of change provides a more inclusive view than invoking "industrialization," but the impulse to find a word or phrase that encapsulates developments is equally problematic here. On the lighter side, we are stuck with cumbersome phrases: "industrializing America" has to give way to "the rise of American market society" or "commercializing America." The issue of description versus explanation seems less grievous, but as with our use of *industrialization,* the question still can be raised, why did the society become market-oriented? The real trouble, however, lies with the descriptive merits of the notion. The explanation that the great economic and social transformations of the nineteenth century in the United States were due to the emergence of the market is based on the premise that America before 1800 was premarket or precapitalistic. That assumption does not quite jibe with the historical record. Stressing the role of the market also overlooks the persistence of nonmarket activities, behaviors, and beliefs into the late nineteenth century in different parts of the country and among various groups within the population. A more important consideration is whether the simple descriptive "market society" is sufficient? Did America evolve into a particular kind of market society? And does the term help us in understanding concomitant political developments?

The chapters in this text are linked by an overall perspective, but one that cannot be conveniently reduced to a simple word or phrase. Instead, we shall examine a more comprehensive way of understanding the great economic and social changes that occurred in this country between 1800 and 1900 than is affordable with such notions as "industrialization" and "commercialization." To state the view taken here as succinctly as possible: developments in the nineteenth century were marked by America's passing first from a mercantile to an unregulated and then to a corporately and state-administered market society.

In greater detail, the argument is as follows: By the turn of the nineteenth century, the Europeans who had settled in the country were hardly a premodern people who had been untouched by market activity. The country began as an outpost of the British mercantile system: the colonies existed simultaneously to build the wealth of the British nation and the power of the British crown. This very mercantilism, however, unleashed forces—physical dispersion, aggressive commercial activity, individualistic behavior—that spelled the doom of that system, and the American Revolution put a belated and formal end to mercantilism. The subsequent path of the new nation was open; America could become neomercantilist with state-driven economic expansion, an unregulated and uninhibited market society, or a more self-regulated yeoman producer republic. For a good hundred years after the American Revolution, politics would be marked by conflict over these vying visions and possibilities. During this time, the undoing of mercantilism gave way to fuller market activity; likewise, continuing antimercantilist sentiment and the democratic institutions established in the late eighteenth century limited government interventions in the economy and fostered a competitive, pluralistic politics. Such forces as population expansion, immigration, and westward settlement promoted (in more important ways) unregulated market activity and political pluralism. The spread of the market spurred industrial development (and urbanization), but neither the growth of market activity nor industrialization occurred evenly or within a vacuum; both processes were shaped by ongoing political dialogues on the future course of the republic. Unbridled market activity created economic and social instabilities and unrest; and by the late nineteenth century all groups within the society—businessmen, farmers, workers, professionals, government officials—began separately to organize and engage in associational activity aimed at the undoing of the competitive economic and political order. The appearance of large-scale corporations in the late nineteenth century represented a greater threat to the ideals of Americans than the earlier emergence of markets. A convergence of efforts at stabilization brought a reformed America by the first decades of the twentieth century.

In this overarching perspective, industrialization is treated as first a product and then an agent of change; industrial development was spawned by increased market activity, but in turn, industrialization both spurred more market activity and created social problems which demanded superintendency of politics and the economy. In this view, change between 1801 and 1901 is thus portrayed not as change from a nonmarket to a market-based America, but rather from an ordered to an unregulated and then administered market society. As will be shown, Americans in the nineteenth century remained divided not over the market per se, but rather, over the kind of market society they wanted, and relatedly, over the nature of social relations and power arrangements within their communities.

As with other general conceptions, the view informing the pages to follow misses a great deal—the United States is too large and its people too diverse for one perspective to incorporate all experience. The portrait provided here will be painted with a broad brush to cover the larger story, but will also be limned with a pointed stylus to highlight the details, the exceptions, and the varying histories of men and women, whites and blacks, native-born and immigrants, the well-off and the left-behind, and people of different regions. A sweeping perspective and narrative also tends to depersonalize the flow of history: vague forces—such as the market, population expansion, and technology—are at work pushing developments. This text, despite the grand themes, will try to put people at center stage—acting with great impact, within limits, with unintended consequences, in opposition to one another, and wittingly or unwittingly in concert.

The discussion to this point is undeniably but perhaps unavoidably abstract. Enough abstraction. Now to the story of the great transformations of American society in the nineteenth century (which will offer evidence of the larger perspective). The first chapter provides an extended survey of social and economic life at the time of Thomas Jefferson's inauguration and treats a revealing contemporary dialogue on manufacture. The second chapter charts the nation's first steps toward industrialization, emphasizing various paths taken and the unevenness of development; various arguments on the causes of industrial expansion will be assessed. Chapter 3 looks at the varied responses of Americans to the changes wrought by early industrialization, and Chapter 4 examines the role of government in economic affairs prior to and during the Civil War. The fifth chapter describes the building of a vast American industrial heartland in the last decades of the nineteenth century, raising questions of continuity and change. Chapter 6 analyzes the rise of the large-scale corporation during the period, and the last chapter deals with the explosive reactions of Americans to the presence of the corporation in the midst of their republic. The building of a new political economic order at the turn of the twentieth century is described briefly at the end of Chapter 7. We begin now in the late eighteenth century.

Industrializing America

Chapter One

Context

Regional Diversity and the Changing Political Economic Order

THE EPISODE had all the trappings of a modern-day spy thriller. In 1787, Tench Coxe, ardent advocate of American manufactures, conspired with a recent English emigrant to have him return to his native country and secure models of the latest textile machinery developed in the famed machine shop of Richard Arkwright. The plan called for the prototypes to be sent to France, where none other than Thomas Jefferson, then ambassador to France, would arrange for their subsequent transport to the United States. To Coxe's dismay, the scheme failed at the start. British authorities caught wind of the plot, and the models were seized by custom officials. British law at the time explicitly prohibited the export of machines, models, and mechanical drawings as well as the emigration of skilled men who possessed knowledge of machine design and creation. The prized secrets of the new industrial age were to be safely guarded at home. Tench Coxe, however, remained undeterred; within a year's time, he apparently had in hand the models and machines he desired.

Tench Coxe was a leading spokesmen for a small cohort of prominent Americans who, during the last decades of the eighteenth century, championed the new cause of manufacture. Among other activities, these advocates founded so-called local Societies for the Encouragement of Manufacture and the Useful Arts, published pamphlets on behalf of greater industrialization, filled the press with correspondence and opinion pieces on the subject, collared new emigrants from Britain for information about the latest developments in machine technology, and arranged furtively for the import of new mechanical inventions. They also attempted in some instances to establish manufactories, largely in textile production, but none of these early ventures survived, falling victim to credit shortages, poor management, and skilled labor scarcity and unrest.

Despite their failures, the advocacy and experiments in manufacture of Tench Coxe and his allies did not go unnoticed or unopposed. Their efforts, in fact, spawned a vital and revealing dialogue. Americans began to discuss the implications of industrial development, and all the anxieties and vying visions they held at the dawning of their new nation came to the surface. This late-eighteenth-century debate on manufacture, which ensued in anticipation of the future, decades before Americans had any real experience with industrialization, lends insight both into the concerns of the day and the nature of this society.

Characterizing the United States in the period immediately following the American Revolution has provoked vociferous debates among scholars. The new nation has variously been described as premodern and capitalistic, and Americans have been labelled traditionalist, communitarian, republican, liberal, and market- and profit-driven. Predominant and prevailing attitudes and behaviors are at issue. Formulating a fixed or certain description for the new nation, however, is problematic, because circumstances and pursuits varied dramatically. Every economic activity known to humanity could be spied, from the hunting and gathering of the original peoples of the continent to the subsistence farming, market agriculture, fishing, timbering, mining, crafting, simple manufacture, and commerce of its settlers. Every kind of arrangement for organizing production and mobilizing labor also existed, from the ancient practice of slavery to the modern wage labor system and the family mode of production and systems based on mutual obligations and service in between. A traveler passing through this country in the late eighteenth century, with proper guidebook in hand, would have had a difficult time at journey's end in affixing a general label to the whole. Social and economic life varied greatly both between and within the different regions of the young nation. However, a complete tour, emphasizing the history and prospects of the different regions, combined with a reading of the contemporary debates on manufacture, does provide a means for reaching a greater understanding. Take that trip starting in the far north of the United States at the time of Thomas Jefferson's first inauguration.

A Tour of the New Republic

New England

Small-scale, family-owned and -operated farms dotted and dominated the New England landscape. With an exception or two, plantation or tenant agriculture failed to take root in Vermont, New Hampshire, Massachusetts, Rhode Island, and Connecticut. (Maine, part of Massachusetts, would not become a state until 1820.) Farm families could be found living in New England in some of the original English settlements of the area, where land had been distributed in a collective fashion by community elders; in newer, closely knit communities established because of religious schisms or population pressure; and in western sections, where lands were bought and sold in less administered ways. Even here, families clustered in definable communities.

The absence of large-scale farm units in New England can be explained by both intentions and circumstance. Settlers to the northern British colonies came not as adventurers, enterprise-builders, or seekers of large profit, or with such prior experience; they came to establish prosperous but ordered and simple communities. Cultural disposition thus mattered. But perhaps more important were the conditions the settlers faced. The particular climate, topology, and soil quality of the region and the absence of east-west navigable waterways blunted any attempts at expansive, staple-crop, commercialized farming. Values, political arrangements, and the environment combined to make for New England's particular early history.

Poverty or subsistence production did not mark the region. This the traveler could easily see. New England farm families worked extremely hard and achieved gains in productivity, but not to produce great surpluses for sale in the marketplace. Family survival, the customary and legal authority of the male heads of households, and the desire to pass on competent legacies in land to the male heirs motivated the labor of the household, not aggrandizement. A division of work by age and sex contributed to the efficiency of the family group. Men generally raised crops and livestock, chopped wood, and fashioned tools and furniture, while women stored and prepared food, made and mended clothes, wove cloth, bore and reared children, tended gardens, and worked in the fields at planting and harvesting times. Girls attended to housework, spinning, and butter making; boys fed the livestock, led plow teams, and assisted in the gathering and storing of crops.

Farm families in New England produced beyond subsistence and a vast range of goods, but they did not become involved in market activity in any but a tangential way. Sometimes circumstances, such as poor transportation and distances to trading centers, prevented market participation, but in most instances, self-sufficiency was a matter of choice. At least 75 percent of the goods produced by the typical New England household in the late eighteenth century went into direct family consumption. Families bartered the rest locally for products and services they could not provide for themselves. Farm women in New England had their own exchange system, often bartering garden crops or midwifery assistance for cloth and other items. If families sold and purchased items in market exchanges with merchants, they did so for luxury goods, such as tea, sugar, and rum. A French traveler to the region in 1787 aptly caught the way of life:

Instead of money incessantly going backwards and forwards into the same hands, [New Englanders] supply their needs in the countryside by direct exchanges. The tailor and the bootmaker go and do the work of their calling at the house of the farmer who requires it and who, most often, provides the raw material for it and pays for the work in goods. These sorts of exchanges cover many objects; they write down what they give and receive on both sides and at the end of the year they settle, with a very small amount of coin, a large variety of exchanges which would not be done in Europe other than with a considerable quantity of money.

The New England countryside would be transformed, though scholars disagree as to whether changes occurred as early as the 1790s or unfolded in a more definitive way during and after the 1820s. Better transportation links would be forged, connecting hinterland with market centers, drawing farmers into market activity; merchants also penetrated into the countryside by establishing rural stores. With population pressures and increased land costs, farm families worked hard to provide legacies and even began to limit family size, as evidenced in declining fertility rates for the region. They also moved to specialize, selling greater proportions of their produce in the marketplace and, in turn, buying necessities as well as luxuries at the stores. Barter decreased and cash transactions grew. The use of hired labor increased slightly, and a market for labor emerged.

Capitalism would come to the New England countryside—there is a question concerning the actual timing—but the continuities in living and working patterns in New England are noteworthy. Here, more than the other regions of the country, farming remained as much subsistence as market-oriented, but from a comparative perspective, almost stagnant. Circumstances again played a role. New England farmers would not be able to compete with the more fertile-field agriculture opening up in New York State and generally in the transappalachian West during the first decades of the nineteenth century. Hinterland-coastal exchanges still remained problematic, with poor transportation linkages; within a few generations, rural depopulation would mark the region. New Englanders also remained fairly fixed in their ways. Transformations thus cannot be overdrawn, although certainly agriculture became more market-defined and -driven.

The uneven character of change in the region is also notable in changing activities within the family. As families specialized and relied more heavily on the market for basic needs as well as amenities, the disposition of labor within the household changed. Women and older children especially were freed from certain activities, but as the cash requirements of the family increased for purchases, they became involved in work taken into the home on a domestic outwork basis. As travelers to New England in the 1780s noted, few families engaged in putting-out work; by 1830 and 1840, great numbers of women in New England were making buttons, palm leaf hats, and shoes in their homes under contract. In Europe, the domestic outwork system classically preceded the rise of the factory and industrialization; in the New England countryside, it would accompany manufacturing growth and mirror the changing economic fortunes, consumption patterns, and labor allocations of the region's farm families.

New England's uneven economic history can also be illustrated in nonagricultural endeavors. Family farms dominated the area, yet in the late 1700s, the region also boasted a prosperous fishing and lumbering trade. Both were tied to the agriculture regime, however, with farmers often working part-time in these pursuits (there would be greater specialization in the nineteenth century). The region also had famous seaport trading centers—Salem and Boston in Massachusetts and Newport in Rhode Island. Examples of diversity, these centers rose on the transport of the resources of New England's forests, lands, and coasts to England but also to the slave plantation colonies of the South and the Caribbean. Great New England family fortunes, in fact, were accumulated through participation in the transatlantic slave trade. Yet, these trading centers declined in the early nineteenth century—Salem and Newport becoming almost ghost towns—as the nation's commerce became centered in New York and other ports. New England changed—became capitalistic—yet, as in agriculture, the story is one of relative stagnation. And the fate of New England agriculture and commerce were intertwined. The region did not have a rich and dynamic hinterland or transportation links that would sustain mercantile success. Again, the story is complex—of the growth of a market economy with the persistence of traditional family-oriented farming, yet also with a plateau reached in commerce and agriculture. Oddly, relative stagnation in these two sectors allowed for phenomenal developments in another, and that is industry. The area would emerge in the first half of the nineteenth century as a center for large-scale industry. Men from families who had accumulated great fortunes in commerce and who now faced declining prospects would bankroll industrial development. And rural young women and children, whose labor was now underutilized, would serve as the initial workforce for New England's manufactories.

Our traveler through New England early in the nineteenth century would thus have found diversity of activity and harbingers of change. Yet, the small-scale family-owned and -operated farm—mostly subsistence and partially market-oriented—would dominate the view. Journeying further south into the Middle Atlantic states a different and even more variegated scene is revealed.

The Middle Atlantic

Self-sufficient farms composed a good portion of the countryside of New York, New Jersey, Pennsylvania, and Delaware, but the traveler's eye would have been caught after leaving New England by the large-scale commercialized farming enterprises that thrived in the Middle Atlantic states. Also noticeable would have been the iron manufactories of the region—which already accounted for 15 percent of world production of iron—and the bustling port cities of New York and Philadelphia, sites of intense commerce, ship building, housing construction, and artisanal production. Tenant farming in large manorial holdings also found a place in the region, particularly in upstate New York and southeastern Pennsylvania. Farms were generally scattered about the countryside, however, not in clustered communities. Properties changed hands frequently, and speculation in land also marked the economic life of the area. The differences with New England would have been striking.

Intention and circumstance must be called upon again to explain the distinct character of the Middle Atlantic states. Dutch authorities during the early colonial period had been more interested in creating trading zones than in colonization (with the exception of the famed patroon estates of the upper Hudson River valley), and this left a legacy for settlement patterns in New York; Moravian and Mennonite groups established tightly knit spiritual communities in Pennsylvania, but the Quakers of Philadelphia and elsewhere possessed a more individualistic and entrepreneurial bent. The area received such a diverse population of immigrants—Dutch, Huguenots, Swedes, Scotch-Irish, Germans, and English and Welsh Quakers—that to a certain extent, the region's varied social and economic life rested on the wide mixture of its peoples. Yet, the area's physical attributes played as important, if not more important, a role. A level and rolling countryside, fertile soils, long growing seasons, and navigable waterways connecting hinterland with seaport cities made for the commercialized agriculture (and the intense economic activity in general) of the region.

A market-driven agriculture, concentrating in corn, wheat, rye, oats, and barley, increased the value of land in the countryside of the Middle Atlantic states, and land speculation and high property turnover followed. Commercialized farming also led to different kinds of productive relations and uses of labor than in New England. Family members worked their homesteads in similar fashion and with similar sex- and age-related divisions of labor, but production for the market required greater land and labor. Slavery and indentured servitude figured in the area's history, institutions almost absent in New England, although slavery never became as embedded in the region as it did south of Delaware and Pennsylvania.

The agricultural fortunes of the Middle Atlantic states thus rested on bonded labor as much as on the energies of household members. Until its formal abolition in the era of the American Revolution, indentured servitude represented a prime labor source. Upwards of 50 percent of the white men and women who immigrated to these shores between the early 1600s and the mid-1770s arrived as indentured servants, and the Middle Atlantic colonies and states received the greatest numbers. Recent research has revealed a great deal about these immigrants who bound themselves to labor for four- to seven-year periods as payment for their transport to the New World.

The origins of the system of indentured servitude are clouded in the historical record. The formal binding of young men for long periods to learn trades as apprentices dates to the late Middle Ages and the creation of the medieval guilds, but service under contract without obligation for trade education had no precedent in England or elsewhere. Indentured servitude may have evolved from the practice of hiring rural laborers on a yearly basis at agricultural fairs, or it may have emerged abruptly out of the extensive labor needs of the British settlements of North America. Whatever its origins, in the 1600s more than 300,000 people signed indentures promising to labor up to seven years in exchange for their passage to the colonies. Although they came, surprisingly, from all walks of life, the servants were predominantly male (only one-fourth were females), young, and poor. The flow of servants tended to rise and fall with economic conditions in England, and starting in the late seventeenth century, the numbers dropped, and the terms upon which servants entered into contract improved correspondingly. Resort to slaves increased accordingly both in the Middle Atlantic and southern colonies. Still, at the time of the American Revolution, the presence of an indentured servant in the Middle Atlantic farm household was common. As with other forms of labor, controlling servants in a circumstance of an abundance of land and a scarcity of hands proved problematic, and while many worked hard and were rewarded at the end of their contracts with bonuses to help them on their way, newspapers of the day were filled with advertisements announcing rewards for the return of servants who had absconded and reneged on their indentures.

A high demand for labor and the use of servants were signs of the intense agricultural activity of the Middle Atlantic states. A prosperous hinterland made for a thriving commerce, and urban development in the region would also have been of note to the traveler. Especially impressive was the burgeoning port city of Philadelphia, center for the intellectual, cultural, and political life of the new republic and, by the late eighteenth century, second only to London in the value and volume of the goods that passed through its harbor. Like New York City, Philadelphia presented a bustling scene. A nascent proletariat of common day laborers worked about the docks in the hauling and carting of goods. Young apprentices and journeymen served and learned under master ship and house builders and in artisan shops fashioning custom clothes, hats, boots, shoes, and furniture for members of the urban elite. A ruder form of shopwork was also apparent at the turn of the nineteenth century: new shops where wage workers produced coarse goods on a more assembly-line-like basis. Orchestrating all of this activity, finally, was a large and diverse community of merchants.

With the products of the Middle Atlantic countryside in demand, particularly in southern areas where plantation slaves toiled exclusively on staple crops and raised little for their own subsistence, there was no dearth of enterprising young men to arrange for the processing, transport, and exchange of exports for imports. The costs of entry into merchandising stayed low through the late eighteenth century, credit remained available from established mercantile houses, and a 12 percent return on average could be expected on investments; as a result, no fewer than one in ten men in a city such as Philadelphia during the period called themselves merchants. They formed a diverse group, from large-scale operators who were members of venerable English trading families to upstart wholesalers and retailers such as the Scottish immigrant Robert Henderson, who made his living shipping flour from Pennsylvania to Charleston, South Carolina, in exchange for indigo and rice. There were profits to be made in such specialization, in developing specific contacts and concentrating in the trade of particular commodities.

Not only had a prosperous and interrelated commerce and agriculture brought prosperity to the Middle Atlantic states, but the region also had been fortunate in escaping the endless imperial wars fought in Canada and New England by the French and British which had brought personal suffering and periodic economic and financial crises to that more northerly region throughout the 1700s. Prosperity and equality, however, were not synonymous, for social divisions in the Middle Atlantic could not but be noticeable. In Philadelphia, 15 percent of the population in the late eighteenth century toiled as unfree laborers, as slaves and indentured servants, and another 30 percent lived in poverty or at its edges. The top 10 percent of the city's taxpayers, on the other hand, owned no less than 90 percent of the taxable property. Inequality marked the region and was more visible than in New England.

The Chesapeake

Social distinctions loomed even larger and more evidently south of Pennsylvania and Delaware in the Chesapeake Bay region. The scene in general in Maryland and Virginia, which included some of the oldest colonial settlements in North America, could not have been more different from New England and the Middle Atlantic states. A warmer climate year long, flat lands along the coast, an area with extensive natural waterways and irrigation—the physical landscape stood as a mark. But the built environment also rendered this a different world. Huge plantations with opulent mansion houses and scores of African slaves working the fields dominated the traveler's view. If the tourist searched diligently, the diversity of the Chesapeake—with tenant farming along the coast and subsistence agriculture in the backcountry—would be revealed, but the formidable plantations of Maryland and Virginia left a lasting impression.

Tobacco—and a craze for smoking in Europe first among the aristocracy and later, as productivity increased and the price declined, among Europeans at large—drove the Chesapeake's early history and was the source of its great prosperity. The region possessed the perfect environment for the growing of tobacco, demand existed (here, circumstance, not intentions, appears to have played a singular role), and economic activity in the Chesapeake literally rose and fell with fluctuations in the price of the precious weed.

Tobacco culture made for two unique social features of the Chesapeake: slavery and the relative absence of cities and urban life. The reasons for slavery taking hold in the Chesapeake (and elsewhere in North America) persist as an issue of debate among scholars. Initially, tobacco in the region was raised on family-owned and -operated farms with the occasional use of servants. The English historically had little experience with the institution of slavery, and the first Africans who arrived in the region in the early 1600s—in slave ships that had blown off course—were not enslaved but, as with white labor, indentured. An abundance of land, a scarcity of labor and huge demand for a product effectively grown in the region exerted pressure for the development of a management system that would bond labor to the land, but questions emerge. Why ultimately slavery, when indentured servitude could have functioned as well? And why were just black people enslaved?

Historians have found important answers to these questions in political and economic events of the last decades of the seventeenth century, when the first slave codes were written—a great legal revolution with people of dark skins now declared slaves, a condition to be held over the course of their lifetimes and automatically passed on to progeny. White fear or hostility toward Africans cannot be seen as a prime ingredient in this legal change, for Africans had already worked as indentured servants and later as free farmers in the Chesapeake for two generations. Political conflict between small- and large-scale white property holders over issues of taxation, representation, and western development provides the political backdrop for the first codification of slavery that occurred in Virginia in the 1660s; placing blacks in a substratum position elevated the white yeomanry and defused class conflict among whites. At the same time, prosperous conditions and increased opportunities in England reduced the flow of indentured servants across the Atlantic. As the price of indentured labor increased, large-scale producers of tobacco looked to a cheaper labor source: African slaves, who were increasingly available with the expansion of the slave trade over the course of the century. Whatever the respective roles of class conflict among whites, simple economics, and basic intolerance, slavery became a fixture in the Chesapeake by the 1700s and added to growing social distinctions in the region, with a small elite of plantation owners now lording over a population of black slaves and white tidewater tenant and backcountry subsistence farmers.

The place of slavery distinguished the Chesapeake from New England and the Middle Atlantic states, and so did the absence of cities. The few urban centers that emerged in the region—Annapolis and Baltimore in Maryland and Norfolk in Virginia—paled in comparison to the burgeoning cities of the northern regions. A number of factors combined to limit urban development in the Chesapeake. An extensive river system in the area made the building of large trans-shipment centers unnecessary. Tobacco also required no processing and could be shipped directly from plantation loading-docks to Europe. An efficient use of slave labor also necessitated that slaves work during slack times between planting and harvest seasons. Slaves grew the food they ate and manufactured the clothes they wore and tools they used. In charge of self-sufficient units, tobacco-plantation owners did not have to rely on urban providers of goods and services. The Chesapeake Bay region, without cities and possessed of a staple-crop, slave-based agriculture, thus shared little in common with areas to the north, other than their mutual battle with the English and attempts to form a more perfect union after the Revolution.

The Lower South

The distance between New England and the states of the Lower South, below the Chesapeake, could be measured in more than miles. North and South Carolina and Georgia appeared to the contemporary traveler as almost another country, closer in character to the British settlements in the Caribbean than to the regions to the north. With its climate subtropical and its coastal lands swampy, this region was even made perilous by its exotic diseases. Moreover, the settlers to the Lower South came not to establish stable communities, but in search of marketable staple crops that would make them rich quick. Immigrants from the West Indies—who had found few opportunities in the Caribbean, as large-scale sugar operations dominated the landscape—first settled the Carolinas and made their money trading deerskins with Indian groups in the area. In the eighteenth century, it would be rice, like tobacco in the Chesapeake, that would generate great wealth and development in general. Indigo plants to be processed into blue dyes and tar and pitch from the area's forests also became key exports, but rice dominated.

Rice cultivation fostered a unique kind of social system in the Lower South. The growing and harvesting of rice was labor-intensive, and profitable in large-scale ventures. The Carolinas and Georgia accordingly witnessed the building of the largest plantations in North America with the largest slave labor forces. In coastal areas, black slaves came to greatly outnumber the white population of plantation owners and yeoman farmers. Hired supervisors managed some large units, and the region was the only one with absentee landlords. Many plantation owners spent a greater part of the time living in Charleston, South Carolina, escaping the torpor of the plantation. Rice growing, finally, also required constant labor, and because plantation owners and managers sought the greatest return on their investments, labor power was directed away from the growing of food and the manufacture of clothes, tools, and other necessities. Lower South plantations became importers of goods as a result and provided an important market for the products of the northern colonies and states. In numerous ways, the plantation economy of the Carolinas and Georgia looked dissimilar even to the Chesapeake and closer to the sugar islands of the West Indies.

The late-eighteenth-century traveler could, in fact, book himself or herself on a boat for Barbados or Jamaica to make the comparison. There, he or she would find huge staple-crop plantations, managed by paid overseers for landowners who remained in England, staffed by armies of slaves who worked full-time in sugar production, and importing food and goods, notably from the Middle Atlantic states. The scale of operations may have distinguished the Caribbean plantations from comparable units in the Lower South, but they shared common features. One noticeable difference between the two areas, however, would be the absence on the islands of a yeomanry, who in the Carolinas and Georgia eked out an existence inland from the great coastal rice plantations.

The late-eighteenth-century traveler to the Caribbean would also see a world no longer linked politically to the North America settlements, but vitally connected on an economic basis. In fact, the sugar islands represented a fulcrum point in a transatlantic system of trade comprising England, western Europe, Africa, and the North American colonies and states. The profits made in the trafficking in human bodies—of which the West Indies received great numbers and in which North American merchants participated—were a key source of wealth in this system. The West Indies had also developed as the primary market for the bounty that was farmed, timbered, and fished on the land and in the forests and seas of New England and the Middle Atlantic region. In fact, the general well-being of the people of those areas through at least the 1780s hinged directly on the ability of West Indian plantations to import goods and indirectly on the price that sugar fetched in the world market economy. When the West Indies prospered, so did the residents of Pennsylvania, New York, and Massachusetts. If the diverse states of the new republic shared any feature in common, it was their historic grounding in the British imperial system of trade.

In touring the new United States, a traveler might not apprehend that fact. The regions of the new nation looked too different—as to living patterns, economic activity, and social relations. Travels through the states would have revealed the varied character of the new republic; and the diversity between and within regions would have left the greatest impression. This was a world both market and nonmarket in orientation that, seemingly, no single description could fit. Perhaps from the outside, however, perhaps from seeing the new nation's place in a larger world or at least in an Atlantic setting, the traveler might gain a means of placing this society as a whole. A visit to the West Indies provides such a perspective: that the inhabitants of North America in common were part of a transatlantic exchange system.

This same holistic view is also available in the dialogue stimulated by Tench Coxe and his co-conspirators in boosting the cause of manufacture in the last decades of the eighteenth century. That discussion similarly reveals the greater society as one formed as a mercantile appendage but also now shedding its past and entering an unknown future. A reading of the debates complements our imaginary tour through the new republic.

The Late-Eighteenth-Century Debate on Manufacture

Advocates of manufacture in the late eighteenth century provided a mix of justifications. Benjamin Rush, one of America's great scientific and philosophic thinkers of the age, offered no fewer than five reasons for encouraging industrial development in a speech delivered in 1775 at the inauguration of the United Company of Philadelphia for Promoting American Manufactures: (1) Americans would save money by manufacturing their own goods and reducing imports; (2) science and industry could help improve agriculture; (3) the poor and indigent could be employed in manufacture; (4) immigration, especially of skilled hands, would be encouraged; and (5) the then rebellious colonies would lessen their dependency on England and be less subject to English law and whims. Rush, in one of the earliest pronouncements on the subject, presented the basic arguments annunciated by later proponents with some additions. Two of his points came to dominate the pro-manufacture position. The role that domestic industry could play in making the nation economically strong and independent emerged as one key notion in disquisitions on behalf of manufacture in later decades, particularly in the early part of the nineteenth century. The ability of industry to employ the idle also became a constant refrain, and here Rush announced a point that would be repeated often: women and children, in Rush's and others' eyes, constituted a population disproportionately poor and slothful, and manufacture would make them more honest and industrious. Their labor would also lower the total cost of adopting machinery. Decades before America's first factories would be staffed by women and children, advocates of industry linked manufacturing work and women's and children's labor. Manufacture, one group in Massachusetts declared, promised "to give employment to a great number of persons, especially females who now eat the bread of idleness. . . . Our design [in building a factory] is not to enrich ourselves." Another group pledged to make work "for infirm women and children, who want for employment and are often burdensome to the Public."

Tench Coxe, the most active of advocates for manufacture, took for granted the benefits of industry and spent a good deal of his time refuting skeptics and opponents, but he too addressed the issue of idle labor. A relative scarcity of labor, particularly skilled hands, some argued, would prevent the development of industry. Coxe noted the potential use of women and children and the curbing of their idleness, but argued that the point of employing machinery was to dispense ultimately with the entire need for labor. Coxe also argued that agriculture would not suffer with the spread of industry and the drawing off of resources, but would in fact be stimulated.

The most complete advocacy of manufacture was contained in a report on the subject prepared in 1792 for Congress by Alexander Hamilton, secretary of the Treasury. (Hamilton received Coxe's assistance in writing the document.) Hamilton is often depicted as the chief proponent of manufacture, but his role in the late eighteenth century in encouraging industry was at best tangential. Hamilton had not participated beforehand in the movement on behalf of manufacture; he delayed in submitting his report to Congress; and he had little to say on the issue after 1792. His defense of industry, in fact, had more to do with his desire to bring fiscal stability to the new nation than with promoting industry.

In his report, Hamilton echoed Benjamin Rush and others but added some new justifications. Manufacture would promote the immigration of skilled hands and provide work for the idle. "It is worthy of particular remark," he noted, "that in general, women and children are rendered more useful, and the latter more early useful, by manufacturing establishments, than they would otherwise be." Industry would also generally increase productivity in society and achieve a greater division of labor, in which farmers could concentrate on farming and others focus on providing goods and services for them. He also argued that industry would lessen dependency on foreign supplies and that those employed in manufacture could absorb the surpluses of the land not sold abroad. A domestic market had to be developed where world market conditions were so uncertain. Finally, Hamilton offered two new arguments: if the new nation had to defend itself against military attack, a manufacturing base was vital to produce the implements of war; second, and playing a prominent part in his report, manufacture would open avenues for investment and direct surplus capital toward productive purposes and away from speculation and consumption of foreign luxuries. Stabilizing financial markets—particularly in government securities—and balancing the nation's trade deficit loomed as great ends for Hamilton, and industry represented a critical means.

A defensive tone, though, marked Hamilton's manifesto on behalf of manufacture, as it did the arguments of others. Adherents repeatedly insisted that industry would not harm agriculture, but would promote and assist farming. Industry would deter indolence and indulgence in foreign luxuries—two other refrains. These were not idle points, for late-eighteenth-century proponents of industry could anticipate fierce responses to their proselytizing. The creation of a manufacturing sector in the society—not machinery or technological improvement per se—represented for many a grave threat to democratic republican institutions.

Benjamin Franklin, man of science and invention and certainly no enemy of mechanical innovation, fired some of the earliest salvos against industry. "Manufactures are founded in poverty," he wrote, for "it is the multitude of poor without land in a country, and who must work for others at low wages or starve, that enables undertakers to carry on manufacture." To avoid creating great disparities of wealth and power and the potential for social unrest, Franklin counselled that the country should extend its agricultural base and remain a simple society of independent producers.

Thomas Jefferson and James Madison later expanded on Franklin's warnings. Both leaders despaired at the prior histories of republics. Republics invariably succumbed to despotism. The growth of propertyless masses crowded into cities and hired cheaply without any greater obligations as to their welfare had proved to be the death knell of all previous experiments in republican government. Easily appealed to by dissenting factions, the urban dispossessed became the base on which potential Caesars constructed their power. For Madison, separation of powers and decentralized government—hallmarks of his federalist system—provided one means of foiling despotism. For Madison and Jefferson, another was the fostering of conditions under which virtuous, public-minded citizens would emerge; such conditions required an economic system based on dignified work that fostered autonomy and heartiness—in other words, freehold agriculture.

Characteristically, Jefferson bequeathed the most eloquent statements on behalf of agriculture: "Those who labour on the earth are the chosen people of God, if ever he had a chosen people, whose breasts he has made his peculiar deposit for substantial and genuine virtue." "Cultivators of the earth are the most valuable citizens . . . the most independent, the most virtuous, and they are tied to their country and wedded to its liberty and interest by the most lasting bonds." "While we have land to labor thus, let us never wish to see our citizens occupied at the work-bench, or twirling a distaff. . . . Let our work-shops remain in Europe. The loss by the transportation of commodities across the Atlantic will be made up in happiness and permanence of government. The mobs of great cities add just so much to the support of pure government as sores do to the strength of the human body."

Madison and Jefferson in their orations and writings did not envision a static or simple pastoral republic of subsistence farmers. They cannot be interpreted as being necessarily antiprogress or antimodern. They pictured a prosperous nation of productive farm families who worked hard to provide basic necessities and surpluses to be bartered for other goods made locally or imported from abroad. In anticipation of growing population, they saw the nation expanding decisively across the continent—thus the importance of the Louisiana Purchase and plans to move Indian tribal groups further and further west—and supported trade policies guaranteeing that America's agricultural surpluses could be marketed and exchanged overseas. They did not imagine a passive society. Later, too, they became advocates of manufacture, especially as commerce with Europe and the importation of manufactured goods entangled the nation in European wars. Jefferson, also a man of science, schooled himself in mechanical invention; and he and Madison, in the first decades of the nineteenth century, noted their recognition that manufacture had to be developed to lessen the nation's dependency on foreign supplies of goods and to build a dynamic internally based economy. Yet, both continued to caution against the creation of large-scale enterprises and the employment of massive numbers of wage laborers. They encouraged manufacturing endeavor in households and small to medium-sized workshops—especially as long as opportunities for freehold agriculture existed and social discontent could be alleviated with an expanding frontier.

Franklin's, Jefferson's, and Madison's position against manufacture—an argument against an expanding manufacturing sector, not against machines—must be understood in view of their hopes and fears for the new republic. Avoiding the worst aspects of European society played a critical role in their responses to advocates of industry. When Jefferson and his followers looked to Europe and particularly England, they saw royal despotism, court favor and corruption, aristocratic opulence, the privileging of the merchant community, rural depopulation and degeneration, and urban growth, poverty, and crisis. In Europe, manufacture meant either the great workshops of the crown, which produced luxuries and encouraged venality, or the urban manufactories employing the multitudes of displaced and poor of the society, which included great numbers of women and children. Madison and Jefferson warned that manufacture would exist to produce luxuries and encourage pomposity and self-indulgence. Advocates of manufacture thus took pains to argue that manufacture in the United States would lead only to the production of necessary goods and, in fact, divert attention away from the importation of European baubles. Manufacture, in their eyes, would further the employment of the indigent and spur continued agricultural development.

For Madison and Jefferson, manufacture was only part of a greater evil and that was the mercantile political economic order recently over-thrown—a regime based on crown-orchestrated development, the use of commercial interests to build the wealth of the nation (and indirectly, the coffers of the monarch), ministerial corruption, and the exploitation of colonies. The argument against manufacture then came as part of a critique of mercantilism and defense of the postmercantilist republic that they hoped to establish. Nowhere is the antimercantilist thrust of the critique of industry better exemplified than in the controversy spawned by the creation of the Society to Establish Useful Manufacture (SUM) in 1791.

In the early 1790s, Tench Coxe developed an idea for a larger indus-trial experiment. A corporation would be founded, stock sold, and capi-tal accumulated for the building of an industrial city. Coxe conferred with Alexander Hamilton on the project, and Hamilton rallied a group of New York City investors and close allies to the cause. For Hamilton, SUM represented a prime opportunity to stabilize the chaotic market for public securities. Written into the charter of the new corporation, which was approved by the New Jersey state legislature, was a clause that stock in the company had to be purchased largely with federal bonds and shares of the Bank of the United States; New Jersey legislators also af-forded SUM other privileges, including tax exemptions, power of emi-nent domain, the right to raise money through lotteries, relief from taxes, and immunity from military service for workers recruited to the project. In his report on manufacture to Congress, Hamilton in fact had con-nected the nation's fiscal stability to the establishment of large industrial corporate ventures.

Once chartered, the directors of SUM began plans for a great indus-trial works near the falls of the Passaic River in what would become Paterson, New Jersey. Waterway systems were built to tap the power of the river and construction proceeded on a series of mills where an im-pressive array of goods—including paper, shoes, pottery, textiles, and beer—were to be manufactured. The bankruptcy of key investors, mis-management, and labor problems stopped the grand scheme in its tracks, and within a few years the whole endeavor collapsed. SUM had an im-pact, though, for the project provoked fierce attacks that had repercus-sions in American politics for decades to come.

SUM seemingly embodied the worst aspects of the former mercantile order. Here was a minister of government, one also suspected of monar-chical leanings, conspiring with wealthy associates—mostly merchants and bankers—to create a private venture that received public privileges and a monopoly position. A manufacturing city with masses of wage laborers was to be erected in the pristine wilderness of the new republic.

The government is "planting a Birmingham and Manchester" in our midst, one critic charged, alluding to the infamous British industrial towns. SUM represents "a political monster," another averred, "a great danger to the republican principles which ought to be dear to every American." "Under the pretext of nurturing manufactures, a new field may be opened for favoritism, influence and monopolies," declared a third. The severe opposition to SUM gave proponents of manufacture pause and effectively stilled their movement. Jefferson and his supporters would use SUM and other incidents in the 1790s to build an anti-Hamiltonian and anti-Federalist Party coalition and consensus.

The debate over manufacture in the late eighteenth century and the reaction to Tench Coxe's grand plans for Paterson, New Jersey, provide a means of comprehending the new nation. Diversity marked the American republic at its dawning, and yet there is a way to grasp the whole. Characterizing American society at the time as either traditional or capitalistic, nonmarket- or market-driven, republican or liberal, misses the mark—and not just because of the varied nature of activity and belief. The dialogue over manufacture, like the larger perspective affordable from a trip to the British West Indies, points to the country's mercantile heritage and base.

The North American settlements came into being as mercantile appendages. The new inhabitants of the area—whether in tidewater or backcountry—were all part of a particular kind of commercial system; they were not a premodern people unimpinged upon or touched by market experience. Yet, this was not simply a market society; Americans operated in a specific political economic order that was being shed for an unknown future. The citizens of the new republic were moving away from a society marked by crown orchestration of economic development and empowerment of mercantile groups—all in the name of nation building. The system had been withering away, in fact, because the seeds of the destruction of mercantilism lay within the expansionism and self-seeking behavior it fostered. The British had not yet developed the administrative experience or expertise to manage the far-flung system; and the colonies, through tacit evasion and plain circumstance (distance from the mother country, the openness of the frontier), freed themselves from the regulatory regime. The American Revolution represented an official end, a political recognition of what already existed in social and economic reality.

Yet, despite formal severance from the mercantile system, the United States remained still in the mercantile orbit. The fortunes of the now former colonialists remained in the grips of British commercial interests; the ups and downs of economic activity during and after the 1780s reflected the abilities of British merchants to flood the American market with goods or block American exports. Politically, too, a mercantile position still persisted in the United States in the person of Alexander Hamilton and in the policies of the Federalist Party. Hamilton and his allies sought to build a powerful nation through commerce and through empowering commercial and financial elites. In some sense, the ability of the new nation to emerge totally from its mercantile past came not with the American Revolution, but with the War of 1812, which led to the demise of the Federalist Party and the end of entanglements in British (and European) affairs.

The United States was shedding its mercantile cast in the late eighteenth and early nineteenth centuries. That is how best to characterize the society at the time. What was to be built in its stead was unknown. What would not be built was clear and that was a world of centralized power, governmental favor, monopoly sway, and aristocratic pretension. Americans were caught not between building a nonmarket or a market society, a communitarian or individualistic one, but between a mercantile and nonmercantilist one: that captures the essence of the moment and the character of the new nation. Historians have often pointed to the seeming paradox that Alexander Hamilton was a social conservative and economic modernizer, and Thomas Jefferson a political and social liberal but an economic traditionalist. Those contradictions are lessened when the Hamilton/Jefferson split is seen as mercantilist/antimercantilist divide. Jefferson envisioned his agrarian republic marching and expanding into the future. Once the mercantile option was eliminated, of course, different paths appeared. Was America to be an unregulated republic of virtuous producers or of aggressive accumulators? Or was the process to be more orderly, with new elites ready to guide the march of progress? That remained to be discussed and decided.

Scholarly debates concerned with the nature of American society at the turn of the nineteenth century are mired because attention normally focuses on individual attitudes and behaviors. Establishing whether the majority of Americans were traditional or individualistic, market- or nonmarket-driven is an impossible task. Usually, the statements of elites are invoked to back one interpretation or another; whether those pronouncements actually represent the views of different Americans, be they men or women, old stock or newcomers, whites or blacks, rich or poor, is not known. Diversity of activity and experience is known. Analyzing the whole is complicated by the diversity, yet the notion that American society was transforming from a mercantile order dissolves such alternative perspectives as premodern/modern or nonmarket/market, and a transatlantic-based tour of the new nation as well as the eighteenth-century debates on manufacture are useful in this respect.

Forces would soon be at work which would catapult the new nation into the industrial age. That process unfolded without any grand plan on the order of Tench Coxe's or Alexander Hamilton's and in spite of the antimanufacturing position. Yet, the process did not occur in a vacuum nor was it unimpeded. Industrialization would transpire amid continuing debates about the nature of relations and power in the society, and that context mediated developments. New discussions would also echo many of the aspirations, concerns, and anxieties voiced in America's first dialogue on manufacture, a dialogue that occurred decades before the first real steps were taken toward industrialization, when America was an agrarian society in a larger mercantile fold.

Chapter Two

Paths

The Unevenness of Early Industrial Development

GRIM AND GREY: those are terms that come to mind when we visualize the nineteenth-century industrial city. Mill buildings dominate the vision, smoke spilling from their stacks, clouding the sky, blackening all surfaces. Masses of faceless men, women, and children are also filing through factory gates—it is six o'clock in the morning—taking their places in industry; working twelve hours a day, six days a week; servants to the machinery, feeding and tending it with relentless, repetitive motions; exposed to interminable heat and noise and constant physical danger; paid a pittance for their labors; stumbling home at workday's end to crowded tenements and dismal flats, families barely making ends meet. This is an apocalyptic view, of sudden and complete transformation: the industrial revolution.

The image reflects the reality only in small part. Complexity and diversity marked the whole process of industrialization. The growth of manufacture occurred in different countries, regions within countries, and trades in a remarkably uneven fashion; the timing and pace of change varied widely, as did choices of technology and managerial arrangements. Industrialization destroyed certain skills and occupations but created others; the process similarly generated both small- and large-scale enterprise. In the United States, agricultural expansion, not contraction, accompanied industrial development; professional, clerical, and service-sector employment grew alongside. No blanket history is thus affordable. The complex character of industrial development can be illustrated by first examining four different paths traveled toward industrialization in nineteenth-century America.

Industrialization in the Countryside:
The Mill Village

The notice in the newspaper caught Samuel Slater's eye. In 1787, he was just finishing his apprenticeship as a management trainee in a cotton mill in the village of Milford in Derbyshire, England. The news article attracted his attention because of his recently acquired knowledge in new water-powered cotton textile technologies. He read with interest that one John Hague had been awarded an extraordinary prize of £100 from the Pennsylvania Society for the Encouragement of Manufacture and the Useful Arts for building a successful water-driven carding machine, a device that straightened cotton fibers for spinning quickly and on a high-volume basis. This was probably the same John Hague who had left Derbyshire a decade earlier to find his fortunes in the New World. Slater undoubtedly had heard of other reports of Americans who sought and purchased the services of anyone from Britain who had been privy to pioneering developments in manufacture. The article on Hague convinced Slater that great opportunities existed for him across the Atlantic, and he made quick plans to set sail. He had to take care, however. He disguised himself as a farm laborer before boarding a ship to New York City. Had British customs officials suspected his training and knowledge in industry, he would not have been allowed to embark. (Obviously, the British system of barring emigration by skilled workers had cracks.) Slater thus reached the United States in 1789, and he would soon play a notable role in America's industrial history.

Samuel Slater is the best known of several scores of British skilled mechanics and mill managers who transferred the secrets of the new industrial age to the United States. Americans would greatly benefit from England's early start in manufacture, and the first generations of American manufacturers, particularly in textile production, would rely heavily on British immigrants who brought information and drawings or models of carding devices, spinning frames, power looms, and cloth printing machines. The British textile machine shop of the late eighteenth century functioned as the key site of development; from these bustling places came in quick succession the inventions that ushered in industrialization. A few figures—Richard Arkwright, most notably—served as the chief sources of inspiration and design, but the process was remarkably collaborative, with teams of skilled mechanics devising, constructing, and perfecting the new machinery. These workers and their contributions have passed into obscurity. Yet their place in invention cannot be ignored, and some of these English artisans would also serve in the diffusion of technology to the United States. American industrialization cannot be owed to the British: the great majority of British immigrants arrived with little industrial experience, the relatively few who did could not possibly have met the total need for expertise required by expanding production, and circumstances in America necessitated technological modifications and innovations not contemplated or followed in England. Still, in the early period of American industrial development, English artisans contributed vital knowledge and helped launch a process.

Samuel Slater made sure that his contribution would not be lost to the historical record. Not only would he add to technological development in the United States, but he also would chart a particular path to industrialization in this country. Upon his arrival in America, Slater found employment in a New York City spinning jenny factory. In the then fairly small world of American manufacture, Slater heard of the problems faced by a prominent Providence, Rhode Island, merchant, Moses Brown, in establishing a textile works. Slater wrote to Brown offering to build and manage a spinning mill with Brown's money in return for a hefty percentage of the profits. With the dearth of his kind of expertise, Slater could drive a hard bargain. Brown accepted Slater's conditions, and thus began a collaboration that would see the building of dozens of cotton mills throughout Rhode Island and southeastern Massachusetts, including the construction of Slater's Pawtucket, Rhode Island, mill in 1793, commonly designated as the first successful mechanized spinning operation in the country (the operative word here is "successful").

Slater acted as an agent of technological transfer, but he also helped forge a particular kind of production system. He faced a singular problem in establishing textile manufactories; he could construct buildings and machines but not staff them. Agriculture drew all available labor in the vicinity of his mills. Slater first tried to operate his works with orphans and poor children who had become wards of their town governments, and later with apprentices, but to no avail. He then moved to attract and hire whole families. He entered into contracts with male heads of households. Wives and children would work in the mills; fathers would be offered jobs in supervision, construction, farming on lands purchased by Slater and the Browns, or in full- or part-time weaving in cottages provided by the company. Slater used the patriarchal authority system of the New England household to his own advantage, with the male heads responsible for maintaining their part of the contract—that is, to supply disciplined labor.

Necessity—the scarcity of labor and an abundance of land—forced Slater to rely on a family system of production, but experience and intention played a role, too. Derbyshire mills had operated in the same fashion, and Slater transferred this knowledge to the New England countryside. Slater also chose to build model, harmonious communities about his mills; attracting labor figured in this decision, but so did Slater's particular belief system (and his English experience). The mill villages he helped found typically included the mill at creek or river's edge; scattered cottages for families of workers; a commons and main street with school house, church, and stores; and artisan shops. Fields also about the village were worked normally on a leased basis.

An aura of stability held in Slater's communities. Family values and dynamics, religious teachings, schooling, and managerial rules contributed to the creation of order. Yet, tension and change marked the mill villages as well. Slater was in constant conflict with his investors, who did not care for his paternalist schemes; farmers surrounding the villages vigorously objected to the damming of riverways and diversions of water to power the mills; fathers could discipline their charges, but also serve as their representatives and demand new concessions; families constantly just up and left; formal work stoppages occurred. By the 1830s and 1840s, newcomers unconnected to Slater's operations had begun to open enterprises; and a growing labor market of immigrant and native-born young people who no longer found opportunity in New England's declining agricultural lands allowed Slater and his followers to begin to hire on a simple daily wage basis. Competition, too, from large-scale producers of threads, yarn, and cloth spelled an end to several mills and the communities they sustained.

Numerous circumstances could force the disintegration of the incorporated mill village, yet the form prevailed, spread, and persisted. From the northern wilderness of Maine to the hamlet of Rockdale in southeastern Pennsylvania, along creeks and narrow rivers, sprang up hundreds of mill communities during the first half of the nineteenth century—more than 400 by 1820—some actually founded by former employees of Slater. They shared many characteristics. Prominent local families sponsored development. The services of immigrant and native-born artisans were utilized in the design and construction of buildings and machines. Scientific and technological breakthroughs were followed and greatly discussed. Whole families of workers were employed, although widows and children composed sizable proportions of these communities. Women and children worked directly in the mills, adult men in handloom weaving in cottages and in ancillary positions. Mill village manufacturers concentrated in spinning and the weaving of cloth on an outwork basis. Mill village textile businesses generally survived through sales in local or specialized markets; eventually, many would succumb to competition from large-scale producers. Attempts to found stable communities only partially succeeded: workers bowed out of church services, children skipped schooling, and the benevolent order in general required too much energy to sustain. Families also stayed for short periods, and wage laborers, many of them immigrant, eventually took their place.

As the mill villages of the Northeast began to disappear or lose their singular character in the last decades of the nineteenth century, they began to make their appearance in other sections of the country, particularly in the South. The family labor system of production remained a basic component of American industrialization; so, too, did attempts to create ordered industrial communities in the American countryside. Labor scarcity generally served as the prime motivator in the settlement and employment of whole families in rural industry, but the shared values and expectations of manufacturers and the families lured to the mills contributed to developments. Another kind of enclosed industrial village would also appear in the mid- and late-nineteenth century, particularly in mining and lumbering areas—the notorious company town. Marked by coercive labor practices, little benevolence, and the employment of individuals, not families, these outposts were a far cry from the communities established by Samuel Slater.

The classic mill village of the nineteenth century is normally and rightly associated with textiles. In fact, for all intents and purposes the so-called "industrial revolution" of the late eighteenth and early nineteenth centuries was a revolution in textile production. With the possible exception of the steam engine—whose invention was stimulated by the need to pump water from the deep shafts of coal mines—the great inventions of the times were textile machines. The factory, too, emerged as an institution in the period basically to house carding, spinning, and weaving machines. The new textile industry also served as the base for further manufacturing development—the chemical industry emerged to meet the needs of textile producers for bleaches and dyes; the machine and tool industries arose as similar offshoots. Textiles dominated events in the early period—the next technological surge would be related to transportation developments—yet, progress in the manufacture of other goods did occur in the early era and also in the countryside.

The resources of the lands and forests of the New World had provided for the growth of industry during the colonial period. Sawmills and grain mills dotted the countryside along falling waterways, and the early nineteenth century witnessed both expansion in production and improvements in milling technology. Oliver Evans' multistoried, fully automated flour mill, developed in the late 1780s alongside the Red Clay Creek in Delaware, represented one of the most spectacular leaps of invention (and imagination) of the whole early industrial period. Rural ironworks, taking advantage of rich iron ore deposits and the vast woodlands of the

American countryside, continued to increase the nation's contribution to world iron production; logs burned in special ways provided the charcoal fuel for the rural iron furnaces. Paper mills, requiring enormous quantities of water, were similarly located in the hinterland, and a major expansion in paper manufacture would occur in the 1820s with the adoption of the Fourdrinier machine, which fully mechanized the papermaking process. Production of lumber, grain, iron, and paper thus progressed in the countryside and, in the cases of iron and paper, often in the context of the mill village, though with individual laborers as well as family groups.

Industrialization is usually identified with cities—as it was by critics of manufacture such as Benjamin Franklin and Thomas Jefferson—yet industrialization in the United States (and England and elsewhere) began and continued in the countryside. Falling water to power the new machines made rural areas the site of initial and further manufacturing development; other resources of the land and forest played their role as well in the location of early industry. A special form of organization of production also appeared and proliferated in the countryside: the mill village with the family system of labor. Manufacture in administered rural communities on the basis of the employment of whole families represented one essential path toward industrialization.

Full-Scale Industrialization:
The One-Industry City

Francis Cabot Lowell did not rely on British immigrant artisans to provide him with information on the latest developments in textile technology. He went to England to see for himself. In 1810, Lowell, an established Boston merchant and Harvard College mathematics graduate, left his home for a two-year stay in Britain. He traveled ostensibly for health reasons, but also in search of new investment opportunities. The strain and uncertainties of commerce had taken a toll on his physical well-being, and investment in manufacture appeared more stable and less involving. Lowell visited several textile mills, showing a keen interest in new power looms, and he took mental notes. When he left England in 1812, custom officials searched his luggage, in vain, to find incriminating writing or drawings.

Lowell returned to Boston bent on constructing an integrated spinning and weaving mill that would produce an affordable broadcloth with state-of-the-art machinery. With the assistance of a close friend, Nathan Appleton, a member of another venerable Boston merchant family, Lowell founded a corporation, received a charter from the Massachusetts state legislature, and raised from among other elite families of the area $400,000 to launch his venture (ten times the capital invested by the Browns and Slater in their individual mills).

Lowell and Appleton chose a site for their mill in Waltham, Massachusetts. Lowell may have held the secrets of the power loom in his mind, but he required the services of a brilliant mechanic, Paul Moody, to see its actual construction. The problem of attracting a labor force also concerned Lowell. A teeming and seething industrial city on the order of Manchester, England, had to be avoided; relying on displaced farm families, in the way that Slater had, seemed impracticable given the large staffing needs of his planned spinning and weaving factory. Lowell and Appleton decided on another strategy: to employ young, unmarried women from respectable New England farm families. The women would be provided safe lodging in company-owned boardinghouses overseen by matrons, good cash wages, and wholesome recreational activities. In 1814, the Waltham mill of the Boston Manufacturing Company commenced operations and quickly met all hopes and expectations of investors. Another mill building was erected, and soon the water power available at the location was exhausted. Lowell and his associates then made plans to build a larger industrial works at a site twenty-five miles north of Boston at the grand falls of the Merrimack River. Lowell himself would not live to see the awesome industrial city fashioned there from the 1820s to the 1850s that would bear his name.

Nathan Appleton gathered the financial support of approximately eighty Boston men of wealth and standing for the ambitious undertaking, raising more than $8 million over fifteen years' time. Land was cleared, canals dug to channel water from the Merrimack, and sites leased alongside for the building of mills. Progress ensued. By 1835, twenty-two mills lined the canals of Lowell; by 1855, there were fifty-two, employing 8,800 women and 4,400 men and producing 2.25 million yards of cloth each week. An imposing industrial city had emerged in the wilderness; and travelers from Europe, such as Charles Dickens, flocked to Lowell in the decades of its development to see this now-famed example of American ingenuity and enterprise.

Lowell represented a grand leap in business financial practices, the organizing of production, the application of technology, and the employment of labor. The sheer amount of funds raised for this private endeavor dazzled the contemporary mind. The use of the corporate form of ownership was also unprecedented. Family ownership and partnerships constituted and remained the norm throughout the nineteenth century in commerce and manufacture; firms typically developed and expanded through the investment of family savings, the plowing-back of profits, and the taking on of partners, not with the sale of securities.

The consolidation of production also had no analogue. Under the roofs of Lowell mills, cotton was cleaned, carded, spun, woven, and finished. The four-story factory buildings of the city encapsulated the revolution in integrated manufacturing. Cotton was often prepared for spinning on the top floor of these buildings and spun on the third; bobbins of thread were then taken to the second floor for weaving in power looms; finishing, printing, and packaging occurred on the main floor. The only part of the process not integrated was sales; Lowell mill owners relied on commission sales agents to distribute their bolts of cloth.

The flow of production in the Lowell mills dazzled the visitor, but so did the use of water-powered machinery, particularly the looms. English inventors had perfected power weaving machines in the first decades of the century, but it was Yankee manufacturers who adopted the machinery in a wholesale way that astounded the British. Only a few noted, however, that a surplus population of cheaply employed handloom weavers in Britain made innovation with machinery, the substitution of capital for labor, a less pressing matter there than for producers of cloth in labor-scarce New England.

Finally, beyond the great departures in business practices and technology manifest in Lowell, the city also offered an extraordinary human story. The Yankee farm girls tending the machinery of the new mills, their boarding in company dormitories, the publication of their own literary magazine, the *Lowell Offering*, and the general attention to their moral well-being, drew great interest. The majority of these recruits came from middling, respectable farm families and stayed for but short times in employment, as Francis Cabot Lowell had hoped; but they often came for their own reasons—to escape their families and not just to serve as extra-income producers—and they eventually proved to be thorns in the sides of management. The Lowell boardinghouses, conceived as proper and safe environments for these defenseless maidens, became bases for their own political and trade union organization.

Lowell is often rendered as the epitome of American industrialization: corporate ownership; large-scale, fully integrated mechanized production; manufacture of a standardized good; the use of a cheap labor source. The Boston investors who established Lowell replicated the model in other parts of New England—in such places as Chicopee, Massachusetts, though not always on as grand a scale—yet the Lowell system remained exceptional, just one route to industrialization in the United States. The fanfare over Lowell blinded perception of developments elsewhere. Even among other examples of large-scale industry in the early age of manufacture, Lowell is not representative. A case in point here is another monument of American industry that emerged in New England at the time, Lynn Massachusetts, a famed center of shoe production.

In 1870, Lynn looked liked Lowell: large factories, mechanized production of a standardized product (in this case, shoes), and the employment of a wage-earning labor force of men, women, and children. The structural similarities, however, belied a vastly different history. Lowell emerged out of the proverbial thin air—an American phenomenon. Lynn constituted an evolving story, a progression toward full-scale industrialization common to Europe.

In the late eighteenth century, the area in and around the community of Lynn became a center for shoe production, but on the basis of the domestic outwork system. Farm families received raw materials from merchants, and through a division of labor in the household, shoes were produced in slack times; women sewed the uppers, and men shaped the soles and eventually fastened or lasted heels and soles to the uppers. As demand increased, merchants provided more work, and shoemaking became a full-time pursuit. Typically, Lynn shoemakers built small workshops—so-called "ten-footers"—near their homes and even began to hire apprentices and journeymen. Women continued to stitch uppers in the home without direct compensation.

In the 1830s, a centralization in production occurred. Merchants and enterprising shoemakers faced with great demand recruited labor to large workshops, where manufacture could be better supervised and made more efficient. The ten-footers began to disappear, though not entirely. Women in households—not just in Lynn but in a widening arc through the New England countryside—now received direct orders to stitch uppers from central shop owners. Paid a piece-rate price, their work became less connected to the general family labor system, and their new status led to a separate protest and organizing movement on their part to guarantee a just reward for their labors.

The 1840s witnessed further concentration in production. Central shops expanded into factory buildings. Employers began hiring women directly to sew inside the new manufactories, though outwork on a sizable scale persisted. A final stage of development was reached in the 1850s and 1860s with adoption of sewing machines and other mechanical devices for shoemaking. Lynn thereby provides a classic version of industrialization: from home production to domestic outwork, centralization and increased division of labor, factory building, and last, mechanization. Here, machines appeared as a final stage in the capitalist reorganization of production; in Lowell, mechanization led the way. Lynn and Lowell differed in one other aspect. In Lynn's mechanized factories, tools and the product remained in the hands and command of the cutters, stitchers, and lasters; these workers did not serve as simple machine tenders as in Lowell's textile mills. Not coincidentally, artisanal traditions persisted in Lynn and sustained worker protest in a more notable way through the nineteenth century. Lowell and Lynn can be grouped together as examples of full-scale industrialization, of one-industry cities as well, of a different path from the mill village, yet their distinct histories deserve emphasis and reinforce the notion of the very unevenness of industrial development.

Specialization:
The Diversified Manufacturing Center

Lowell and Lynn provided visible evidence to mid-nineteenth-century foreign visitors of American prowess in manufacture. For the traveler to the nation's major urban centers—New York City and Philadelphia, in particular—the view of industrial development was less clear. Government reports indicated that these burgeoning cities were locations for the country's greatest industrial output. But from whence did this production flow? The metropolitan skyline revealed factory buildings here and there, but nothing on the order of the Lowell mills.

A deliberate reconnaissance would find production flowing everywhere: in cellars and attics, tenement flats, artisan shops, and a proliferation of indistinguishable small and medium-sized manufactories. Getting a handle on the swirl of enterprise required peeling away at layers of activity. Lowell, and even Lynn, offered more straightforward pictures.

Describing industrial growth in America's mid-nineteenth-century metropolitan centers is difficult. There are no famous figures (a Slater or Lowell), no prime-moving trades (textiles), and no singular inventions to anchor the story. Thousands of separate endeavors have to be told. Three tales of enterprise from the city of Philadelphia serve here to represent and illustrate the complex and diverse character of industrialization in the metropolis.

Samuel Wetherill was born in 1739. In his teens, he was apprenticed as a house carpenter, and later as a journeyman, he participated in the building of Philadelphia's first textile mill. In 1784, he changed occupations and opened a retail store, where he sold imported iron ware, window glass, tools, and paint pigments. At this time, Wetherill joined with Tench Coxe and other Philadelphians of standing in becoming a public advocate of domestic manufacture. In his own business, he also moved toward manufacture by beginning to grind lead and mix it with oil and colors to produce paint.

In 1804, Wetherill built a separate facility to increase his output of paint. Growing demand for paint with increased housing construction and Wetherill's staunch support for ending the new nation's reliance on imported goods served as the impetus for this greater move toward manufacture. Six years later he built a larger facility, and his business in paint prospered with the curtailing of imports at the time of the War of 1812. To ensure an adequate supply of raw materials, Wetherill purchased a so-called pig lead farm outside of the city.

Wetherill's heirs continued to expand paint production; in the 1840s, his sons oversaw the construction of a large paint factory that employed seventy-five workers. They also closed their father's retail store to concentrate in manufacture. Until 1933, the Wetherill family continued to produce a paint that locally earned a reputation as a product of the highest quality. The enterprise founded by Samuel Wetherill in the late eighteenth century in many ways epitomizes enterprise in the American city; it was a small to medium-sized family-owned and -operated firm that survived through the production of a specialized good. Samuel Wetherill's passage from merchant to industrialist, however, is less characteristic.

William J. Young provides an example of the more typical artisan-turned-manufacturer of the early industrial period. Young is typical in another odd way. He was born in 1800, but existing documents make it unclear as to whether in Scotland or in Philadelphia soon after his parents migrated there. In one person he represents both the native-born and immigrant enterprising craftsman, figures inherent in the story of urban industrialization.

At the age of thirteen, Young was apprenticed to Thomas Whitney to learn, as his indenture noted, "The Trade or Mystery of a Mathematical Instrument Maker." Whitney taught the young man the precious craft of fashioning surveying devices, and Young was fortunate eventually to inherit his master's business in 1825 on Whitney's death.

With western land speculation and development, the time was ripe for the production of mapping instruments. American surveyors had relied on primitive homemade or imported European measuring tools. During the 1830s, Young patented a number of improved devices and received critical orders from new railroad companies, and soon his surveying gauges were earning prizes and acclaim as the finest manufactured in the country. By the 1850s, his shop was producing more than 150 instruments a year.

Despite increased orders, Young throughout his life operated his business as a sole proprietor and according to strict craft standards. His workforce, never numbering more than twenty, included apprentices and a core of highly skilled machinists (whom Young paid dearly, as there was great incentive for them to leave and become independent instrument makers). Instruments fashioned in Young's shop were made to order and crafted by individual workmen; in the face of competition from assembly-line producers, Young and eventually his heirs resisted engaging in even small-batch orders. Young also became active in radical politics in Phila-delphia, serving as an officer in the Working Men's Party in the early 1830s. There is little surviving testimony from either William Young or Samuel Wetherill to warrant anything more than speculation, but within these two figures can be seen different strands of postmercantilist belief, with Wetherill supportive of a dynamic manufacturing republic and Young of the small producer's democracy.

William H. Horstmann provides a bona fide example of immigrant enterprise. A native of Cassel, Germany, Horstmann emigrated to Phila-delphia in the first decade of the nineteenth century. He arrived in Amer-ica with both talents and resources. He was a highly skilled, French-trained silk weaver who enjoyed the financial support of wealthy Phila-delphia relatives.

In 1815, using capital advanced by family members, Horstmann estab-lished a small shop. With the help of other German-born weavers and younger American-born journeymen and apprentices, his shop turned out high-quality silk labels, ribbons, and threads. To avoid competition with foreign imports, Horstmann deliberately specialized in custom products rather than engaging in the production of broad silk cloths.

In 1824, Horstmann introduced to his shop and to the country the revolutionary Jacquard loom, a loom whose movements were controlled in an automated fashion by hole-punch cards. Production and work re-lations at Horstmann's changed immediately. Horstmann soon replaced his apprentices and journeymen with unskilled machine tenders, largely from Philadelphia's German community; he did retain a core of artisan weavers who continued to work on handlooms to produce the more ornate products.

Following the introduction of the new technology, Horstmann's com-pany prospered, but still through the production of specialty goods. From his expanding shop came silk tassels, lace, fine threads, flags, and banners, among a host of items. The firm soon outgrew its original quarters, and in 1854, Horstmann's sons opened a five-story factory building near the center of Philadelphia; they employed now between 400 and 500 workers, who continued to manufacture fancy silk prod-ucts. Horstmann's operated on a far larger scale than was typical in the mid-nineteenth-century American metropolis—in 1860, manufacturing firms in Philadelphia averaged a mere eight employees—yet, in specialization of production, the company joined its neighbors.

Take the enterprises of Samuel Wetherill, William Young, William Horstmann, and the thousands of others that operated in cities such as New York, Philadelphia, and Newark at mid-nineteenth century, and they do add up to a whole. At least four characteristics are apparent in metropolitan industrialization. The first is product diversity. Instead of one kind of item, an amazing array of goods poured from workshops of the city: paints and varnishes, fine instruments, fancy cloth, hats and caps, plain garments, tailored wear, tools, machines, saws, lumber, furniture, rugs, chemicals, drugs, glass, jewelry, books, bricks and tiles, and more.

A diversity of work settings is a complementary feature. The goods produced in the mid-nineteenth-century American city issued from a variety of sites. Urban workers toiled in factory buildings, operating water-powered, but in some instances, steam-powered machinery; in smaller manufactories with hand- and foot-driven machines; in artisans' shops where craft practices and standards persisted; and in sweatshops and homes. Workplaces constantly changed, too. A visit to a mill building one year might find one firm operating with a force of machine operatives; the next year the same urban structure might be subdivided into several businesses producing variously on manufactory, craft, or sweatshop bases. Single goods could be manufactured in any number of the city's diverse workplaces—hats and caps, for example, were produced in factories as well craft shops—and, adding to the complexity, single goods might pass through several settings in the complete production process. In textiles, a fiber thus might be combed and carded in a home on an outwork basis, spun in a mill, woven in an artisan shop, and dyed or printed in a small manufactory. The line between mill, shop, and home was hazy.

Specialization—in both operations and products—was a third component of the urban production system. Fully integrated enterprises on the order of the Lowell mills were more the exception than the rule; separate establishments emerged as the pattern. In textiles, for example, independent manufacturers engaged in either spinning or the weaving, dyeing, or printing of cloth. Custom production also marked urban industry. Rather than produce coarse standardized goods, city firms prospered by manufacturing small-batch custom items to the specifications of their many clients; this occurred even in large-scale works such as William Horstmann's silk company.

The small to medium-sized family-owned and -managed business was a fourth critical feature of metropolitan industrialization. As late as 1880 in a city like Philadelphia, the average industrial worker labored in a unit of approximately twenty employees; the number of large firms on the order of those common to one-industry towns, with 750 or more employed, could be counted on the fingers of one hand. Only a small percentage of companies adopted the corporate form of ownership—proprietorships and partnerships were the norm—and the few corporate entities that emerged in major urban centers were owned privately and founded for legal reasons and the privileges of limited liability rather than for greater capitalization.

Diversified products and work settings, segmented establishments, a high proportion of employment devoted to the production of specialized goods, and the prevalence of the small to medium-sized family-owned and -administered enterprises characterized the mid-nineteenth century urban industrial system. Why did the structure of urban production take this form? Why was it so different from the one-industry mill towns? There is no single or simple answer here. Energy resources is a first consideration. Cities such as New York and Philadelphia lacked major running waterways or waterfalls; this initially precluded the building of large, totally mechanized factories. Urban firms would eventually look to solve their energy needs with steam power, but an adequate supply of labor would support the persistence of hand labor.

Artisanal production also provided the base for the ever-steady increases in urban industry. What is notable in America's major cities in the nineteenth century is not the effort of merchants gathering outworkers into merchant-created centralized shops and manufactories, but the building and expansion of firms by craftsmen-entrepreneurs. The artisans-turned-manufacturers also fortuitously had on hand plentiful supplies of skilled laborers, allowing for the production of fine wares. Cities continued to attract skilled immigrant workers and potential small-scale operators, perpetuating the process.

The advantages of specialized production were a third consideration. Competing with large-scale manufacturers—from New England and eventually elsewhere—would have proven disastrous for urban proprietors. Rather than produce standardized goods, they profited by dealing in local or specialized markets. Their small scale afforded a flexibility that allowed them to shift into new product lines with fashion and market changes; abundant skilled labor further facilitated the process. The urban production system rested on the advantages achieved through specialization rather than the benefits that can accrue through scale.

A final partial explanation for the particular industrial history of America's major nineteenth-century cities lies in the investment behaviors of urban elites. The great mercantile families that had accumulated wealth in colonial commerce placed their surplus capital in large-scale ventures: in Lowell mills, but also in further trade, banking, canal and railroad construction, and mining development. They opted out of urban industrial development. This created a capital scarcity, a further limit on the building of large enterprises, but also a vacuum. Aspiring native-born artisans and enterprising immigrant skilled workers filled the void and established the diversified urban manufacturing system and center.

The Southern Variant: Industrial Slavery

The antebellum South is not identified with industrialization, but rather with the expansion of slave plantation agriculture. Cotton is the story here. Cotton had been grown before the 1800s, but southern economic development through the colonial period had been based on the cultivation of tobacco and rice (as well as indigo and the processing of pine pitch and hemp for naval stores). A leap in southern fortunes occurred in the early nineteenth century with an intense shift toward cotton production. Eli Whitney's invention of the cotton gin—a device patented in 1793 to separate cotton seeds from the cotton fiber, particularly short-staple cotton—traditionally is cited as the prime mover. The cotton gin developed by Whitney and others eliminated a time-consuming and labor-intensive process, thereby making the growing of cotton a profitable pursuit. But the real source of the South's new and prosperous history was the British textile revolution. The South possessed the optimum soil and climate to grow what was now the world's main commodity: the cotton demanded by the new machines. Industrialization in this case generated greater agricultural development.

Cotton production dominated southern history in the first half of the nineteenth century, forcing population dispersion, land speculation and western development in the region, the perpetuation and encasing of slavery, and eventually political conflict between the states. Yet, the period also witnessed manufacturing growth in the South. The South had its advocates of industry, who experimented with textile machines and mills in the late eighteenth century and formed their societies to promote the cause. In the 1830s and 1840s, another set of proselytizers emerged to encourage Southerners to invest in local industry, arguing that the region could ill afford to depend on outsiders, namely Northerners, for needed manufactured goods.

And growth occurred. By 1860, the South accounted for 20 percent of the capital invested in the nation's industries and 15 percent of the country's manufacturing capacity. Twenty-five percent of the textiles produced in the United States at the time came from southern mills, although this and other statistics have to be put in proper perspective: southern textile factories combined contained fewer spindles than found in the city of Lowell alone. The South lagged behind the North in general; with 36 percent of the American population, the region contributed less than 11 percent of the country's industrial output. Yet, if the South had become a separate nation, it would have then ranked in the top five or six of industrial nations, and not just in textiles, but iron making, mining, milling of grains and timbers, sugar refining, and leather tanning as well. To the South's disadvantage, too, lost from the accounting of production was the vast amount of manufactured goods produced on plantations for direct use.

That the South developed industries at all might have come as a surprise to a contemporary visitor who expected all energies to be directed toward cotton. Equally of note was the antebellum South's system of industrial production, for actually laboring in the region's manufactories were slaves. The choice of labor systems had provoked debate among southern proponents of industrialization. William Gregg, a chief advocate, warned against the use of slaves. Slaves remained too inefficient and undisciplined for industrial work, he argued. Southern manufactures would be better advised to both utilize and improve the labor of the white poor of the region. To that end, Gregg established a mill village on order of Samuel Slater's in Graniteville, South Carolina, in the 1840s, where he employed white families from backwoods areas. William Gregg became the exception, however, for southern industrialists in the antebellum period would rely almost exclusively on slave labor. In the 1850s, between 150,000 and 200,000 slaves, 5 percent of the total number of slaves in the South, toiled in southern textile mills, iron works, tobacco processing plants, hemp factories, sugar refineries, coal mines, salt works, grain and lumber mills, and in construction and on the region's railroads. If the number of slave artisans on plantations—carpenters, blacksmiths, and others—were added to the list, the ranks of the region's slave industrial labor force would be more impressive.

Slaves working in industry often possessed greater skills than field hands on the plantation, yet their labor could be more grueling. Manufacturing firms in the South generally purchased their own slaves; 80 percent of the slaves engaged in manufacture were owned directly, and the other 20 percent were hired, on loan, in effect, from local slave owners. Companies provided clothing, barracklike housing, food, and health care. The industrial regimen led to long hours and constant toil, close supervision, injuries, and no time off; the irregular pattern of agricultural work afforded greater leeway. Industrial slaves rebelled in the face of such exploitation, collectively with work stoppages and sabotage—often led by the skilled workers whose craft sensibilities additionally fired their anger—and individually as runaways. Hired slaves were spared some of the worst brutalities; their owners signed contracts with industrial employers that provided a modicum of protection from severe punishment.

Slavery thus did not preclude manufacture; the South charted its peculiar path to industrialization with the use of bonded labor. A question, though, concerning antebellum developments has occupied scholars: the South witnessed the growth of manufacturing during the first half of the nineteenth century, but why did the region lag so far behind the North? Why was there not greater industrialization in the South during the period? To these questions a variety of answers have been offered.

The very success of agriculture stilled industrial progress. Great profits were to be made in investments in land, slaves, and cotton seed, given the great demand for southern cotton in the world market economy. Few chose to put their money elsewhere. The South also lacked a viable market for manufactured goods; most plantations were self-sufficient or importers of cheap items from the North and abroad, and the white yeomanry were either too poor or resistant to participate in market activity.

Slavery also served as a brake on manufacturing growth. Slaves could never form a large industrial labor force, for slave masters feared concentrations of slaves in industrial sites, particularly in urban areas. The creation of a free wage-labor force, particularly of immigrants, was equally threatening. A final explanation for the southern lag in industrial development places weight on the conservatism of the planter elites, their lack of an entrepreneurial bent.

None of the above conjectures provides a satisfactory answer. Southern industrial enterprises returned great profits, two and three times the returns on investments in agriculture in most instances; more capital should have been attracted to manufacture. The South bought manufactured goods from the North and Europe; local industries could have supplied this market. Obviously, too, slavery could be adapted to industrial pursuits. The South thus had during the antebellum period ample incentives and resources for greater manufacturing progress.

A key element in the question of southern industrialization has to be the proclivities of the men of wealth who could have bankrolled further developments. Those who accumulated surplus capital through agriculture either felt the risk too great, feared the social consequences of establishing manufacturing centers, or else felt no pressure to change their activities. Merchant capitalists in the North eagerly sought new outlets for investment with the strain and uncertainties of commerce; they funded large-scale projects, and enterprising native-born and immigrant artisans filled in the voids. The power and authority of the elites of the South rested on their lordship of the land; they were not pushed economically to move in new directions, and they acted to preserve the basis of their status and the social order. There were no other groups to seize the industrial initiative. The South, again, wended its particular way.

The Varied Course and Causes
of Industrialization

There is, then, no single history of industrialization. Obvious differences can be noted between nations, but even within one country, as is the case with the United States in the first half of the nineteenth century, industrial development assumed various forms: the mill village with the family system of labor; the one-industry town; the diversified urban manufacturing center; industrial slavery. The disparate character of industrialization has not, however, prevented a search for singular answers, for locating key factors, causes, or prime movers of change. In the 1950s, to provide a notable example, scholars from a variety of disciplines—economics, political science, history—devoted great effort toward conceiving succinct explanations or models of industrialization. The Cold War provided the immediate backdrop to this rather agitated intellectual endeavor. The United States and the Soviet Union at the time vied for the loyalties of new nations created through decolonialization—the ending, after World War II, of centuries of European imperialism. Advice and programs were offered to modernize the countries of Africa, Asia, and Latin America that were deemed undeveloped; in the name of increased productivity and higher standards of living, industrialization emerged as a central goal. To counter the Soviet example and model—government-directed socialist development—American scholars rushed forward with alternative strategies drawn seemingly from America's successful past.

Concentrate on the production and marketing of a single staple crop or raw material that can be produced at a comparative advantage, went one answer. Use the monies gained from sales of that commodity in the world market to finance industrial development. Referred to here was the assumed prime role that cotton had played in antebellum America; the income the South derived from cotton fostered economic development in the Northeast and West as the South purchased goods and services from those regions. Create an infrastructure of transportation and communications facilities to allow and encourage market activity, went another proposal of the period that harked back to other developments in the early nineteenth century. Or place great energies and money into a prime industry—railroads or steel, for example—and through various spin-offs, that initiative would lead to expansion in all sectors. Build schools, increase literacy, and inculcate modern attitudes—a further answer drawn from the American experience that had its adherents.

The scramble to define and advance so-called noncommunist recipes for development can be faulted on any number of grounds (in fact, this entire intellectual enterprise had a short history, though not without great repercussions for American foreign policy). Circumstances differed so entirely that lessons drawn from the 1820s and 1830s had little practical value in the mid-twentieth century. Americans certainly did not face a world of fully developed and competitive countries at the moment of its economic ascendance, for example. Deliberation and plan also mark modern initiatives at development; they have followed revolutions and wars that established new nations. While some advocates of manufacture also portrayed American industrialization as part of a nation-building process, manufacturing emerged in the United States neither orchestrated nor part of a grand scheme. Certain answers likewise represented incomplete readings of the American past. Industrialization in the Northeast, for example, can hardly be attributed to demand for products from the cotton-rich South, as one Cold War–era theory of development suggested; industrialization rested more critically on factors internal to the region, including growing local demand for manufactured goods. Particular criticisms of the theories of economic development conceived in the 1950s can thus be noted, but a major point may be missed: that the entire effort to conceive of recipes rested on the dubious assumption that there is one road toward industrialization.

The varied nature of industrial development renders any discussion of the causes of industrialization necessarily complex. Two general approaches to explaining the transition to industrial society are apparent in the literature on the subject. The first is to focus on the role of *events*. For example, emphasis can be placed on the impact of a major invention or technological breakthrough. Thus, the invention of the steam engine (or the spinning frame) suddenly and dramatically altered the course of history. One could as well pinpoint the trailblazing initiatives of entrepreneurial figures and their investment in and establishment of new endeavors. This way of accounting for change has never proved adequate or satisfactory. Invention (or capitalization) rarely occurs in a vacuum, but rather in response to demands for products that are marks of changes already afoot in societies; likewise, the adoptions of inventions—and the reasons behind them—are as important as, if not more important than, the inventions themselves.

Other kinds of events can also be made the center of the story. America's early industrial history has often been related to European political crises occurring during the first decades of the nineteenth century. The economic fate of the new republic hinged to a great extent on the ongoing Napoleonic Wars in Europe. America's commercial fortunes prospered when American ships remained free from attack and were the sole source of international transportation services; the business of American merchants collapsed when their ships lost their neutral status and came under fire. In 1807, France and England declared all ships carrying goods to the other side to be liable to attack; President Thomas Jefferson responded with the Embargo Act of 1807, which virtually ended all foreign trade of American shippers. The mercantile community in the United States suffered dearly—one reason why Francis Cabot Lowell and others sought new investment outlets—but ironically, suspension of overseas commerce served as a great boost to American manufacturers, who operated now without foreign competition. The seeds of American industrialization in this traditional way of analysis can thus be laid to geopolitical events of the first decades of the nineteenth century, and particularly to the Embargo Act of 1807. As with other explanations that fix on the importance of particular events, this too sidesteps a whole complex of developments.

A different tack than attributing industrialization to this or that invention, individual, or political crisis is to focus on long-term impersonal *forces* or the preconditions and givens that allowed for change. The new republic's vast natural resource base, trading experience and traditions, expanding population (and relatedly, expanding demand for products), supportive political framework, artisanal base, and relative antitraditionalism and unboundedness—all of these served as critical factors. Capital, labor, and ideas could flow toward new opportunities unimpeded for the most part by custom, law, group ties, and the need of elites to maintain social ways (on these latter points the north and south do diverge). The ingredients were there, in other words, for industrialization.

Isolating necessary conditions for change does beg to a certain extent the question of the causes of industrialization. Scholars operating within this explanatory mode often have to resort to some precipitating event or development; once in motion, history unfolds rapidly given the various auspicious circumstances. This structural approach also ignores (or fails to account for) the very uneven and disparate character of industrial development. (The same could be said for perspectives that rest on the role of specific occurrences.) The only conclusion to be drawn is that the causes of industrialization are complex. Underlying forces and critical events have to be considered in tandem; but more important, no discussion of the causes of industrial development can lose sight of its varied character. There is no single recipe.

If different paths were traveled—if the *course* of industrialization varied—the questions arises, then why? Values and political crises and ideology played some role. The potential existed for greater investment and progress in industry in the South, even with the great gains to be made in agriculture. But fears of social disorder with either a slave or free-labor (immigrant) industrial workforce, and the ruling elite's desire to maintain their status as lords of the land, placed a brake on development. In the North, too, concern about social unrest shaped choices. Samuel Slater and the men who launched large-scale industrial projects in Lowell and elsewhere mobilized labor and organized production in ways they hoped would sustain social harmonies. Native-born and immigrant artisans created enterprises that maintained craft traditions and practices. Moral visions and not just economic calculations guided these ventures and made for different histories.

Costs mattered, though, and particular resource endowments and costs contributed to differences in development. Results could be seemingly at odds. Consider wood as an example. The vast forests of the North American continent provided colonists and citizens of the new republic alike with an abundance of wood. Ships could be built cheaply, and the shipping industry thrived in northeastern cities. Inexpensive wood allowed machine builders to fabricate cheap machines, and machine use spread accordingly. Americans quickly learned disposable habits; with wood machines in relatively great supply, manufacturers replaced these less-than-durable constructions at will. In deforested England, in contrast, where wood was scarce, machines were built of wrought iron. They lasted longer, but their cost (and an English taste for quality) slowed the adoption of mechanical means of production. Yet, the very abundance of wood in the United States also led to a retardation of development in one key industry, and that was iron manufacture. Iron makers in the country continued to melt iron ore with cheap carbonized wood (charcoal) into the second half of the nineteenth century; in Great Britain, manufacturers at an early date experimented with treated coal (coke) as a fuel and were producing quantities of high-grade steel in specially developed blast furnaces decades before their American counterparts. A resource endowment did not necessarily induce technological progress; the cost factor could drive innovation or sustain old practices.

An abundance of wood also allowed for cheap building construction;

but here, the high cost of labor—specifically skilled labor—countered expansion. American builders soon developed new techniques—including balloon-frame housing, which required few hands—to alleviate the problem. Labor costs proved during the antebellum period to be an important element shaping developments. The many paths followed toward industrialization in the era can be related in great measure to the labor factor. In labor-scarce New England, manufacturers moved to substitute capital for labor and thereby established the fully integrated, mechanized textile mills of Lowell (they also made use of the region's underutilized labor of women and children as semi- and unskilled machine tenders); Samuel Slater dealt with the problem by employing whole families, who were often displaced from the land. In New York and Philadelphia, a relative wealth of skilled labor allowed for the maintenance and expansion of a special small-batch, custom production system. In the South, the costs of shifting labor fully and profitably employed on the land to industry deterred greater investment in manufacture. The labor factor does not singly explain differences in development. The investment behaviors of Boston's great commercial families have to be taken into account as well as labor costs in the New England region to understand the extraordinary history of Lowell; likewise, the fears and predilections of southern elite planters have to be weighed alongside the expense they might have incurred in substantially moving slave labor from the land to the mill. Still, labor was a resource whose varied cost contributed to a varied history of industrialization.

The question of labor costs and American industrial development has captured the attention of successive generations of commentators and scholars. As early as the 1840s and 1850s, British visitors to the United States focused on the relative scarcity of skilled labor in the country and suggested that the high costs of such labor had driven American manufacturers to substitute capital or machinery for labor in a dramatic way. They were equally impressed with the adoption by American businessmen of interchangeable parts production techniques, which they dubbed "The American System of Manufactures." The fitting of imprecisely tooled components into final products involved time-consuming and highly skilled labor; by improving parts manufacture, American industrialists, according to British investigators, had reduced the assembly process to a simple and cheaply paid task. A revolution in production had thus been achieved.

In highlighting seemingly special features of American industrialization, British visitors bequeathed a definitive view: that America's industrial development had a singular quality, was unique, and that high labor costs were the key generative force. That perspective continues to inform studies of America's industrial past, and in our own time economists still debate what is called the "scarcity of labor" thesis—a beguiling mono-causal argument that can explain not only technological developments, but also, supposedly, the ease with which Americans accepted machin-ery since jobs were not threatened, Americans' penchant for cheap stan-dardized products, the purchasing power of American consumers and American consumerism, the robustness of the American economy in gen-eral, and the relative absence of class tensions in the society. The issue of labor costs is critical in understanding early American industrializa-tion (and in accounting for different paths taken), but qualifications are in order for this singular explanation as well.

First, British visitors saw and wrote of many things. They viewed Lowell and were impressed by the full-scale mechanization of produc-tion. The Americans had adopted their power loom technologies in a rapid and widespread way, and in this instance, the persistence of hand-loom weaving in England, with that nation's abundant supply of weavers in need of work, made a perfectly clear point. The British also made note of the fully integrated production system of the Lowell mills and drew similar lessons about the role of labor costs. But the visitors viewed another kind of work site where they were inspired in a different way. Wooden and brass clock manufactories, located largely in Connecticut, provided one impressive display. There, semiskilled workers stamped, carved, or lathed hundreds of simple parts and passed them along on a conveyor-line basis for the rapid assembly of uniform clocks. This was a far cry from the traditional clock works in which a few craftsmen pains-takingly made custom clocks, often with one-of-a-kind clock pieces. Equally of note was the production of guns in American arms factories and especially in federal armories in Springfield, Massachusetts, and Harpers Ferry, Virginia. Eli Whitney had actually been one of the first manufacturers to experiment with the mass production of guns using interchangeable parts; Whitney is etched into history for his invention of the cotton gin, but he played an equally important role in devising new manufacturing techniques with guns. The great breakthroughs occurred in the federal armories in the first three decades of the nineteenth century with the development of new lathing and drilling machines that required little skill to operate and that produced precisely tooled parts easily as-sembled into guns on a mass-production basis; as late as 1800, all guns, their stocks, barrels, and locks, had been crafted by single gunsmiths. British visitors could not but be impressed by what they saw in the more notable clock and gun works that they toured, and it was the inter-changeable parts manufacturing technique that they dubbed "The Amer-ican System of Manufactures." It was not the fully mechanized and integrated production at Lowell that they had in mind when invoking this cachet.

Unfortunately, British visitors saw only select sites and did not investigate deeply where they did tour; consequently, they handed down an incomplete and misleading view. Had they visited Slater's mill villages or, more notably, the labyrinthine world of metropolitan industrialization of a New York or Philadelphia (much less the manufactories of the slave-labor South), they would not have drawn a monolithic portrait. America had many systems of manufacture, and diversity should have been the key point of their reports, not the exceptional cases of Lowell and the interchangeable parts production clock and gun works they visited. Even at the latter, the British commentators exaggerated the extent to which an assembly-line system had been perfected. (They also missed the labor unrest that marked these places, particularly at the armory at Harpers Ferry.) Knowledge and technology were still in a primitive stage, and late into the century, skilled craftsmen were still filing and bending and getting the parts into place for the proper operation of the clocks and guns produced (and later, the sewing machines and other devices where interchangeable parts production techniques were adopted). Until a truly national and standardized market came into existence as well, there existed little pressure to have the absolutely perfect tooled parts that would make the promise of assembly-line manufacture a reality.

The British saw a piece of the whole, and that extraordinary piece cannot be discounted. Nor can the role that labor costs played in shaping American industrialization, particularly the diverse character of industrial development in the country. High skilled labor costs in New England caused the substitution of capital for labor and the adoption of detailed division and deskilling of labor techniques; a surplus of skilled labor, on the other hand, allowed for specialized manufacture in the nation's urban centers. The "scarcity of labor" thesis itself only holds in certain instances, but a general labor cost view that points to both insufficiencies and adequacies of labor is helpful in understanding the varied nature of American industrialization.

Values, fears of social unrest, and relative resource endowments can be employed to explain the different routes to industrial development traveled in the United States in the first half of the nineteenth century. But two additional considerations deserve mention. The first of these is the very absence of plan: a nation-builder on the order of a Bismarck or a Stalin did not force industrialization in this country; no person or group charted or had the power to fix a particular course, and varied paths could be followed. Second, the nation was too large, locally oriented, diverse demographically and in natural resources, and politically decentralized for there to be one history written. Unevenness naturally marked the process. Early American industrialization, then, had a disparate character, but this did not in fact prevent the American people from reacting in both organized and personal ways to the vast social changes that occurred in the country in the early stages of development. But since a uniform industrial experience did not prevail, a unified response was not in the offing.

Chapter Three

Reactions

Americans' Responses to
Early Industrialization

ON THE EVENING of June 1, 1824, fire destroyed a portion of Walcott's Mill in Pawtucket, Rhode Island, and as a local newspaper reported, arson was suspected. This had not been the first fire of suspicious origin to damage a cotton mill in the seemingly peaceful industrial village developed by Samuel Slater, but the circumstances surrounding the blaze at Walcott's gave extra pause to textile manufacturers in the area. Just six days earlier, workers in the cotton mills of Pawtucket had walked off their jobs in protest over the simultaneous announcements of reductions in wages and increases in the hours of work. The fire on June 1 may have persuaded mill owners in Pawtucket of the deep-seated anger of their employees and the potential for greater disorder, for in two days' time they agreed to a compromise settlement of the dispute, and production resumed.

Burning factory buildings and destroying machinery represented one response to industrialization, but it was not a common one in the United States during the early years of industrial development. Evidence exists for only occasional and scattered incidents of such sabotage. Certainly, the country did not witness the contagion of machine-breaking that gripped England in 1811 and 1812, when manufacturing workers marching under the banner of the mythical figure Ned Ludd destroyed more than a thousand textile mills and hacked spinning jennies, power looms, and hosiery, lace, and shearing frames to pieces. Nor were there any outbreaks of violence equivalent to those accompanying the raids of English agricultural workers in the 1830s, this movement led by a mythical Captain Swing, that resulted in the burning of barns and massive destruction of threshing machines.

The English machine-breakers of the early nineteenth century left many a legacy, including a tangible image of labor violence, but also a label, Luddite. To judge modern technology as inherently dehumanizing and exploitative is to risk being called a Luddite. The term, however, provides a misreading of the events of the early nineteenth century. English workers who destroyed looms and threshers were not antagonistic to machinery per se. The historical record reveals their raids to have been organized, aimed at specific employers, part of strike activity, and launched during particularly bad economic times. Machine-breaking was more tactic than principle, a pre-trade-union way of protesting deteriorating conditions of work and life that echoed centuries-long traditions of plebeian revolt. This way of understanding violent responses in England to industrialization in the early period also offers a critical clue as to the relative absence of machine-wrecking in the United States in the same era. The commercialization of agriculture and open-field farming had released labor from the land and created a surplus labor population in England; the adoption of labor-saving devices made matters worse. English laboring people attacked the very objects that caused greater unemployment and powerlessness. In the United States, in contrast, machines replaced few workers; with an expanding agricultural base and with labor therefore lured to the land, machines filled a vacuum.

The machine did not emerge as a phantom in the midst of the new American republic, as a threat necessarily either to livelihoods or social order. Americans greeted the machines of the age with genuine curiosity and enthusiasm. Men like Thomas Jefferson, who greatly feared the personal and political impact of industrialization, assiduously followed the latest technological developments and delighted in their own inventions. Working people and manufacturers alike during the antebellum period flocked to demonstrations of the newest devices and attended lectures and forums on science and engineering at newly founded institutions, such as the aptly named Franklin Institute for the Promotion of the Mechanic Arts in Philadelphia. A Thoreau or a Melville might have depicted the machine as intruder and despoiler of nature, but a more common scene represented in the art and literature of the day was of the machine naturally absorbed into the pastoral American landscape. The machine further appeared in the essays of Emerson, the orations of Daniel Webster, and the poetry of Walt Whitman as a vital tool for the virtuous American yeomanry in whose hands rested the well-being and progress of the community. Artists and writers thus incorporated the machine into their visions of a bucolic American republic, but in the very taming of technology with their pictures and words there rested an admission of technology's potential threat to nature and the society. The example of England hovered constantly in the mind. Still, the positive heralding of the machine had a basis in reality; industrial development by mid-nineteenth century was uneven and at an early stage, and in the American circumstance of a relative abundance of land and scarcity of labor, the machine did not displace.

Americans did react during the antebellum period in protest and fear over transformations occurring in their midst, but the machine was not the problem. At issue instead were fundamental changes in the nature of social relations and community life wrought by the expansion of un-bridled market activity and the spread of the wage labor system. Americans responded in both organized and personal ways to these challenges of the day. Craftsmen protested the undoing of traditional practices in their trades, and workers in emerging industries, the exploitative character of new kinds of work. People of property and standing formed institutions to reorder their fractured communities. A more subtle response involved the assuming and fashioning of new lifestyles and poses. The various reactions to be delineated can be placed under the convenient title of "responses to early industrialization," but that caption requires clarification. Americans responded rather generously to the coming of the machine; they reacted apprehensively to a new age seemingly ruled by supply and demand and the cash nexus. The sum of the reactions, however, totaled less than an explosion, in contrast to social crises that would occur in the United States in the late nineteenth century. The unevenness of development in the antebellum period and varied industrial experiences moderated the reaction; an antimercantilist consensus also held events in check, and in general, the stage had not yet been reached where industrialization fundamentally challenged the groundworks of the republic.

Artisan Protest

Change and conflict materialized first in the artisan shops of the new republic. Mechanization and factory production, hallmarks of industrialization, played only a small role here; at issue, rather, was the more fundamental question of the nature of social relationships in the craft shop. The craft system can be traced back to ancient times, but modern craft practices date to the creation of the guilds in thirteenth-century Europe. Artisans in medieval cities banded into associations that received special privileges from state authorities. The guilds obtained control of local entrance to trades; in return, they accepted responsibility for the upholding of standards of production and the regulated training and employment of labor.

The guilds began to disintegrate in Europe in the sixteenth century under the pressure of expanded commerce. In England, the Crown moved to bolster the threatened craft system through legislation. Parliament passed the Statute of Artificers in 1562 making guild practices national law. Artisans had to abide by government edicts that controlled the quality and price of goods manufactured as well as the respective obligations of masters, apprentices, and journeymen.

British settlers in the Western Hemisphere transported British laws and traditions with them. The open environment of the New World hampered attempts at establishing guilds and state regulation of trades, production, and employment; but by the late eighteenth century, craft shops in American cities operated by and large as idealized. Master printers, carpenters, silversmiths, bakers, and tailors produced items to order; engaged apprentices and journeymen for fixed periods to serve under them; and remained responsible for their lodging, feeding, and instruction in the so-called mysteries of trades and facilitating their entries into masterhood.

Increased market activity and demand for manufactured products at the turn of the nineteenth century forced change in the American craft shop. The initial agents of change were enterprising artisans and merchants who consolidated the putting-out system into centralized workshops. These entrepreneurs opted to produce standardized goods. They affected detailed divisions of labor in their shops and hired workers on a daily wage basis without any greater obligations to them than to pay for specific tasks completed; labor was thus translated from a fixed to a variable cost.

In the venerable printshops of America, the training of all-around printers who set and inked the type, turned the presses, and even occasionally wrote the copy, came to an end. The trade divided into newspaper, book, journal, and miscellaneous publishing. Setting type became the separate job of compositors, and cheap, nonapprenticed labor was now employed to tend the increasingly automated presses. Craftsmen in tailor shops similarly were turned into either cutters of fabric or finishers of garments; manufacturers of apparel now contracted the sewing of precut materials into clothes to proliferating sweatshop operators in the city. Master cabinetmakers at the same time stopped taking apprentices and journeymen into their homes and shops, where they would be taught all the stages of fabricating fine furniture; hands now were hired to lathe and cut uniform pieces of wood quickly to be fit together by semiskilled workers.

The transformation of craft work in the United States in the first half of the nineteenth century occurred without pattern. Artisanal production persisted in different trades and places, particularly where demand for custom goods existed and where market activity remained localized. Skills also were not automatically eliminated, just narrowed. Printshop owners and garment manufacturers still required typesetters and cutters and finishers; where once young men entered into service broadly to learn skills with the goal of establishing their own craft establishments, apprenticeship programs now prepared workers to occupy specific positions—skilled ones, to be sure—simply as employees. The craft shop also became tied to a larger world of production. Artisan shops received orders from major manufacturers for custom items; they thus thrived as subcontractors, but as in the case of garments, craft shops themselves in turn contracted out work.

Complexity may have marked the transformation of the craft system, but general changes in relations, expectations, and patterns of work produced a visible and vocal reaction in the shop. Journeymen mounted the charge. As early as 1768, a group of twenty journeymen tailors in New York City left their benches in protest over a reduction in wages announced by their masters. Although there are earlier references to job actions in colonial newspapers, this is probably the first strike—or "turn out," as it was then called—launched in America. Printers similarly struck in New York in 1778, seamen in Philadelphia in 1779, shoemakers in New York in 1785, printers in Philadelphia in 1786, and carpenters there in 1791. These work stoppages represented isolated, short-lived events and did not involve any organizations that approximated trade unions.

Relations between masters, their apprentices, and especially their journeymen remained harmonious for the most part during the last decades of the eighteenth century. Men of the shop, in fact, joined in patriotic groups before and during the American Revolution to support independence and the establishment of democratic institutions. They marched together also to encourage the writing and ratification of the Constitution; a national government that could erect protective tariffs and stimulate manufacture was in the common interests of masters and shop workers. They formed organizations, such as the General Society of Mechanics and Tradesmen in New York and the Association of Tradesmen and Manufacturers of Boston, to further promote their shared economic and political aims; these organizations also operated as social, educational, and mutual benefit or insurance societies. Finally, masters, apprentices, and journeymen in general joined at the turn of the century to boost the political party movement of Thomas Jefferson. They upheld the vision of a small producers' republic; they opposed government dispensation of favor, aristocratic rule, and powerful groups such as merchants and bankers who made fortunes not through tangible labor but by hoarding resources and controlling markets.

Nevertheless, the reorganization of production in the shop opened a widening breach between masters and their charges. In the first decade of the nineteenth century, journeymen protested not only wage reductions, but also the lengthening of the work day, the derogation of apprenticeship, general deskilling, and the growing employment of common day laborers. They protested too in a more organized fashion. Journeymen printers, cordwainers, tailors, carpenters, cabinetmakers, shipwrights, coopers, millwrights, stonecutters, handloom weavers, and hatters in New York, Philadelphia, Boston, Baltimore, Albany, Washington, Pittsburgh, and even New Orleans either transformed their old fraternal societies or formed new organizations to demand and bargain for improvements in working conditions. The Federal Society of Journeymen Cordwainers, established in Philadelphia in 1794, led the way; formed as a mutual benefit group, the association soon evolved into the nation's first bona fide trade union and conducted the first organized strike of American workingmen in 1799. The Journeymen Cordwainers in Philadelphia seven years later would also be embroiled in the first great legal trial in the United States involving the rights of union workers. Society members were indicted and found guilty of conspiracy under common law of concerted action to injure others and restrain trade. The right of workers to organize long remained a contested issue, until federal protections were extended to them during the 1930s, yet the decision against the cordwainers in 1806 and others to follow did not stop the further formation of trade unions during the antebellum period, the use of strikes, or a certain amount of leniency allowed by local judges and juries. The nascent trade union movement of craft journeymen also faced more important economic than legal obstacles. Few of the early unions survived the fluctuations in the economy during the second decade of the new century that accompanied geopolitical conflict and a major business collapse in 1819.

A second, broader and more notable surge of protest activity by craft workers would occur in the late 1820s and early 1830s as business revived and pressure mounted anew to change old ways of production. The rupture between masters and journeymen grew. Once again, craftsmen from Philadelphia, and cordwainers in particular, led developments. In 1827, William Heighton, a shoemaker by trade, helped found in the city the Mechanics Union of Trade Associations, the nation's first federated body of unions, and the *Mechanics' Free Press*, official organ of the society and the nation's first labor newspaper. The new association grew out of a strike by carpenters who had organized on behalf of the ten-hour workday; cooperation among workers from different building trades during the dispute convinced Heighton of the power of united action. Heighton used the *Mechanics' Free Press* to defend labor protest and advance the notions that workers should be rewarded all the proceeds received from the sales of the products of their labors and that access to educational opportunities and land should be equal for all citizens of the republic. Heighton further attacked merchants and others who profited not from the producing goods of value, but from taking advantage of market circumstances and receiving privileges from the state. As to the latter, Heighton argued that workers needed to control government; to that end, he called for and helped establish the Working Men's Party of Philadelphia, the nation's first labor party.

Word of these events in the Quaker City spread quickly, and soon Working Men's parties began to appear in a dozen states and scores of localities. Brunswick, Maine; Palmyra, New York; Carlisle, Pennsylvania; and Zanesville, Ohio, among other towns and cities, would soon host local labor parties. A common program is discernible. The parties called for the creation of free, common school educational systems so that children of laboring families could receive the advantages of schooling and not in the setting of the stigmatized community pauper schools. They favored abolition of imprisonment for debt; prohibition of licensed monopolies; reorganization of local militia systems (where recruitment practices placed a heavy burden on young men from working-class backgrounds); legal protections for unions; the abolition of prison labor contracting; and district systems of election. They also called for payment of wages in hard currency rather than in scrip or other unreliable paper notes; passage of mechanics' lien laws to assure that workers had first claims on employers' assets in the event of bankruptcies; and provision of better public services for residents of poor neighborhoods.

The Working Men's parties flourished, garnering sizable votes, and placed labor candidates in office, but the movement peaked early. By 1832, few of the local labor parties remained in operation, most victims of internal factionalism, poor financing, and infiltration by mainstream politicians and absorption into their organizations. However, the demise of the Working Men's parties did not spell an end to protest by craftsmen during the era. Trade union officials picked up the pieces of their failed political crusade and embarked on an intense period of union organizing. Political mobilization spurred their efforts, but so did deteriorating economic and social circumstances. Rising prices, wage cuts, the lengthening of the work day, and further disregard for craft practices and denigration of the journeyman's labor heightened grievances. In every urban community in the United States between 1832 and 1836, craft workers joined together in protest and to defend their interests and status at the workplace.

In Philadelphia, the cordwainers recoordinated their efforts; and hand-loom weavers, bricklayers, plumbers, blacksmiths, cigar makers, and comb makers formed new unions. In New York, cabinetmakers, hat finishers, basket makers, and locksmiths, among others, followed the established examples of printers, cordwainers, and tailors. Baltimore, Pittsburgh, and Louisville likewise witnessed the unionization of bootmakers, stonecutters, coopers, and carpet weavers. Cooperation among workers in different trades also marked this era of union-building. By 1836 central labor councils or federations were established in thirteen manufacturing centers from Boston to Washington, D.C. and west to Cincinnati; in metropolitan areas such as New York and Philadelphia, they were composed of more than fifty separate trade union affiliates. The so-called general trades unions of the period established labor newspapers; sponsored lectures, dinners, and parades on behalf of labor; and most important, fostered interunion assistance during job actions, including the initiation of several general strikes. The most successful of these occurred in Philadelphia in 1835, when approximately 20,000 workers across a dozen trades walked off their jobs to achieve a ten-hour workday and succeeded in having their demands met by their respective employers.

The Philadelphia general strike of 1835 represents a high, and, as it emerged, an end point, for the burgeoning trade union movement of the 1830s. By the end of the decade, few traces of union effort or strength could be found. Economic crisis—this time a long-term business collapse starting in 1837—again decimated union organizations as union members and other workers lost their jobs. Counteroffensives by employers, a restricted ideology on the part of labor leaders, and divisions among laboring people along the lines of skill, gender, race, and ethnicity also contributed to the disintegration.

The organizing of craft workers in the first decades of the nineteenth-century represents one critical reaction of Americans to the changes wrought by industrialization. However, historians have differed in their interpretations of these early protest movements, particularly with regard to the Working Men's parties and subsequent trade union drives of the late 1820s and early 1830s. The authenticity of these latter movements has been questioned. For example, were the Working Men's parties really organizations formed by and for working people? There is evidence of infiltration and manipulation by middle-class reformers and mainstream politicians. What of the message of labor protest at the time? Was it distinctive? Reformist? Or radical? Did the movement succeed? Why did it dissipate?

The following points are offered to provide some perspective on the protests of craft workers during the late 1820s and early 1830s. First, the ubiquitousness and reach of the movement have to be appreciated. Working Men's parties appeared in cities and towns across New England, the Middle Atlantic states, and into Ohio. Community after community in the North was touched by the insurgency. The movement also spawned an estimated twenty labor newspapers that popularized the cause, and at least fifty daily journals in fifteen states reported approvingly on activities of the Working Men's parties between 1828 and 1832. Trade union organizing after 1832 had as wide a spread and sway, with upwards of 300,000 workers enrolled under the union banner during the period.

Second, the movement brought to the fore, and was guided by, a remarkable and articulate group of leaders from within the community of craftsmen. Representative of these leaders were William Heighton, shoemaker, founder of the nation's first local trade union federation, labor newspaper, and labor party; Thomas Skidmore, machinist, controversial leader of the Working Men's Party of New York, who criticized existing divisions of property and called for an end to the inheritance of wealth; John Ferral, handloom weaver, who helped forge a united trade union movement in Philadelphia, led the 1835 general strike, and pointed to political institutions as the source of inequalities in society; Seth Luther, journeymen carpenter, who was a key labor figure in New England and leader of the region's ten-hour movement, as well as a fiery orator, particularly on the subject of accumulations of wealth by bankers, merchants, and other nonproducers; and John Commerford, chair maker, a popular and tireless labor movement organizer from New York City who became a staunch spokesman for the equitable distribution of public lands.

Labor leaders from the craft shops were joined by a group of vocal labor advocates who came from more comfortable backgrounds. For example, Robert Dale Owen, along with his father, the early socialist Robert Owen, founded a number of famed cooperative communities, and among other radical ideas proposed the removing of both rich and poor children from their families and their common attendance in state boarding schools. An associate of Owen's, the extraordinary free thinker Fanny Wright, advocated women's rights, easy divorce, the abolition of slavery, and communal living, and delivered stirring speeches describing the plight of working people in the changing America of the 1830s. George Henry Evans, editor of the *Working Man's Advocate*, became the leading voice for land reform. Thomas Brothers, a hat manufacturer, published another leading labor newspaper of the day that repeatedly attacked financiers and monopolists "as the vilest race that ever infested the world."

The presence of middle-class reformers and radicals in the craftsmen's movement of the late 1820s and early 1830s has provided grounds for some scholars to doubt its authenticity. Indeed, the ideas and personalities of these seeming outsiders created divisions that weakened the Working Men's parties, with a number of trade union leaders at the time expressing distrust of these allies and wondering aloud whether political activity too grievously sapped energies from trade union organizing (a difficult question that would endure). The opportunism of certain figures in the movement, such as William English and Ely Moore, men who actually came from the craft shops and used labor protest as a means to establish places for themselves in mainstream politics, has also encouraged cynical views. This delving into the social backgrounds and motives of the leadership, however, may miss a critical point: what is notable about the period is simply the absolute swirl and sweep of labor activity and discussion.

Labor advocates also spoke with a common, highly politicized voice. In their writings and speeches, craft shop organizers and radical intellectuals shared themes. They upheld a labor theory of value: workers were entitled to the entire proceeds received from the sales of the products of their labors. Workers, in John Commerford's words, were the "real producer[s] of all the wealth and luxury possessed by the rich and powerful." Labor leaders, in fact, saved their greatest invective for merchants, bankers, and speculators, who accumulated wealth not through production; thus, Seth Luther denounced those "who toil not . . . but who are nevertheless clothed in purple and fine linen and fare sumptuously every day" because of the labor of others. The spokesmen for labor also attacked concentrations of wealth as inimical to a democracy. Disparities of wealth in this respect did not flow from natural inequalities in ability or merit. Men were equal by nature; political institutions structured by men generated inequities. John Ferral stated this common theme as follows: "The accumulation of the wealth of society in the hands of a few individuals (which has been abstracted from the producers thereof by means of the erroneous customs, usages, and laws of society) is subversive of the rights of men."

Labor leaders from within and outside the craft shop were also in common men of the Enlightenment. They were Deists, believers in the powers of human reason. They applauded the latest inventions, though not the manner of their application. "Under the *present commercial arrangements*," Robert Dale Owen wrote, "machines, the people's legitimate servants, have become their masters . . . labor-saving machinery, *as at present directed and controlled*, works against the poor man." They similarly petitioned for the ten-hour day so that working people would have the time to read and attend lectures to remain informed of the latest political and scientific developments. The creation of public common schools became a key demand and goal; children from working-class families had to be afforded a proper education to gain the essential power that came from knowledge.

The total program of the labor advocates can be interpreted in various ways. The protest movement of craftsmen in the 1820s and '30s can be placed within the mainstream of American politics. The attacks on non-producers and government favor and the upholding of a vision of a small producers' democracy fit well with the ideals of Jefferson and the contemporary populist politics of Andrew Jackson and his Democratic Party followers (the movement became easily absorbed into the latter). With the possible exception of Thomas Skidmore, who called for the confiscation and distribution of inherited property, labor advocates did not challenge the principle of private property, and in fact, property holding of an equitable kind represented for them the base of independent-minded citizenship. Skilled workers also, in organizing to protect their jobs and improve working conditions, contributed (perhaps unwittingly) to the demise of the craft order; they turned journeymanship, a passage into masterhood, into permanent wage holding, cementing the wage labor system. Protesting craftsmen in this way of thinking did not forge a distinctly oppositional movement.

From another standpoint, the Working Men's parties and the trade union drives of the 1820s and 1830s can be seen as exceptional and radical. Industrial development was in an early stage and was uneven; that labor protest emerged in such a full-fledged manner in the United States at the time is remarkable. Workers in America demonstrated in no less a militant, politicized, and widespread a fashion than their counterparts in England and France. Moreover, their protests cannot be judged by modern standards. They offered no socialist critique, but socialism as a theory or as a rallying cry had little history anywhere by then; early socialists, in fact, could be found in the ranks of the American movement, and radical labor spokesmen raised the issue of redistribution of property. Protesting craftsmen certainly did not embrace the individualist ethos or petty-entrepreneurialism of the Jacksonians. They offered a critique of a society ruled by the forces of supply and demand and opposed concentrations of wealth. They were egalitarians and ultrademocrats, the radicals of their day.

These two different portraits of the early American labor movement can be resolved by placing events in a third perspective. Protesting craftsmen of the 1820s and 1830s joined a debate on the character of the new American political economic order. They had immediate grievances and demands, such as the ten-hour workday, but the movement spoke to larger issues about the republic. The old regime based on crown rule, regulated markets, state creation of privilege, and the empowerment of the merchant class—mercantilism, in short—had been shed. Yet, a new political economic system had not been fixed and remained a contested matter. Craftsmen, like other parties to the dialogue, acted more against the old regime than in favor of a specific future. They attacked tangible vestiges of the old order, such as debtors' prisons and elitism, monopoly power, and governmental favor in general. They contributed to ongoing efforts at making a return to the past unacceptable. As to the future, they feared a new order of competitive, self-interested politics and rapacious economic activity, as well as rule and regulation by new industrial elites. Journeymen imagined a self-regulated society of ultimately equal, hard-working people providing for each other's needs and respectful of each other's labors and rights to participation in the affairs of their communities. That mutualistic vision flowed from their craftshop experience and was the journeymen's particular input into the postmercantilist debates of early-nineteenth-century America. Labeling the movement as radical, reformist, or whatever misses this basic dynamic.

In the end, the journeymen failed to build a sustained working class movement. Internal squabbling and divisions among workers by skill, gender, race, ethnicity, and just experience (the latter, a product of uneven economic development) are cited as reasons for the demise of organized labor activity of the 1820s and 1830s. The clobbering blow of the depression that began in 1837 is sufficient explanation alone. The ideology of the craftsmen's movement also has to be taken into consideration. Their critique of the old and their fears and visions of the future fueled grievance and protest activity, but also served to moderate and impede. Labor advocates challenged various inequities, but not property, wage labor, or market society per se; they had a distinct message, but one that could be subsumed under the Jacksonian banner. More notable, their ideology was highly exclusive. They valued craft labor and citizenship, but these were experiences open only to skilled white men. Their vision did not speak to women, African-Americans, immigrants, common day laborers, and factory hands—groups, in fact, whom the journeymen saw as strikebreakers and replacements, threats to their very livelihoods and workplace and societal ideals. In boosting their cause, in striking a pose of worthiness and respectability, the men of the shop and white workingmen in general delineated themselves from these other workers at the margins and in the process contributed to the racial, gender, and ethnic stereotyping and intraclass divisions of the day. The inability or unwillingness of the craft activists to reach through the expanding and diversified work populace would blunt their own efforts.

Protests by Industrial Workers

The new factories of the republic also became sites of agitation during the first decades of the nineteenth century—witness the events of June 1, 1824, in Pawtucket, Rhode Island—but factory workers did not wait for leadership from the urban craftshops. They took matters into their own hands to demonstrate against the deplorable conditions of their work. This history had an important twist: the story of the first protests of American industrial workers is a story of American wage-earning women. Decades before a Francis Cabot Lowell dared to conceive of a manufacturing center on the order of Lowell, Massachusetts, Tench Coxe and others had argued that a reserve army of women and children could serve in the front lines of industry. And by the 1830s women had come to represent more than a third of the nation's entire manufacturing workforce but upwards of 80 percent of those employed in the great textile mills of New England and elsewhere. Women comprised a disproportionate number of the nation's first truly industrialized workers, and they would launch protests of their own making.

The hope had been that harmony would mark the new textile mill villages and cities of the Northeast. Founders of these communities had provided housing, schools, churches, and cultural events as gestures of goodwill; the engendering of loyalty and diligence was also in their minds. Events did not unfold according to design. The boardinghouses of Lowell, for example, intended as wholesome environments, became perfect centers for the organizing of protest. The mills themselves served as further places to build labor solidarity. New recruits to the mills were assigned to experienced hands, who not only initiated them into the work but also into the mores and grievances of the emerging sisterhood of operatives. Reality in the form of changing economic circumstances played final havoc with best-laid plans. Increased production and competition in the 1830s, in particular, forced mill owners to reduce costs. They looked to lowering wages, lengthening the workday, speeding the machinery, and increasing individual work assignments. The moment became ripe for conflict. While incidents of unrest among mill hands were recorded before the 1830s and would continue to affect mill communities across New England and the Middle Atlantic states, the first great outbreak of organized industrial protest would occur in the 1830s, and, as might be predicted, in Lowell.

In mid-February of 1834, rumors of impending wage reductions swept through the mills of Lowell. Faced with rising inventories and a sluggish market, directors of the Lowell manufactories were in fact debating the issue and were about to post notices. Women mill operatives did not require confirmation to begin organizing. They held meetings, circulated petitions, and mobilized support throughout the mill community. Agitation on February 14 reached the point that work came to a halt in some mills. The dismissal of one alleged leader then led to an initial walkout and protest. Soon, eight hundred women workers, one-sixth of the entire female labor force of the city, took to the streets in a kind of massive demonstration that had not been seen in New England since the time of the American Revolution. At a mass outdoor rally that followed, the protesters resolved to remain off their jobs until they received guarantees that wages would not be reduced.

Protesting Lowell mill hands failed to sustain the initiative; in less than a week's time, the turnout collapsed and the women returned either to the mills or to their homes in the New England countryside. The demonstrations nevertheless shocked local supervisors and the Boston directors of the Lowell textile works. The ease with which the women organized caught them by surprise; suspicions existed that a base for organization had been laid down well prior to the protests. Appearances of harmony were deceiving. Equally notable was the rhetoric of the pioneer industrial strikers. Even though they were unskilled mill hands and deprived of full citizenship because of their sex, Lowell's protesters spoke a language akin to the discourse of the men of the shop. "The oppressing hand of avarice would enslave us," a key petition announced. The women sought just wages as just reward for their hard labors and to avoid dependency; as "daughters of freemen," they deserved proper treatment.

Neither fine organization nor rhetoric helped the cause, but the early collapse of the 1834 strike did not still conflict in the nation's model industrial city. In October 1836, women workers in Lowell walked off their jobs again. This time the precipitating event was the announcement of increases in the price of room and board charged by company boardinghouses. Once more, protesting women workers displayed impressive organizing skills; in fact, an actual organization—the Factory Girls' Association, with a membership of 2,500—emerged to coordinate activities. And again, the strikers spoke of their dignity and of rights being trampled by a new tyranny. This time, however, the protests of the Lowell women proved more enduring and successful. Upwards of one-third of the female workforce in the city joined walkouts called in opposition to increased charges, and the pressure was maintained for several weeks. Better economic conditions in 1836 than 1834 helped the cause. Faced with increased orders for cloth, effective organization on the part of workers, and difficulty in recruiting new hands, employers soon surrendered and rescinded orders to raise the price of lodging.

Opportunities for further organization on the part of Lowell's female mill operatives were dashed, however, as was the case with craftsmen, by the onset of the depression of 1837. Mill owners cut back production in response to falling demand and laid off workers. In attempting to remain in business by assigning more machines to individual employees and by speedups, Lowell's textile manufacturers did lay the basis for another outburst of protest by their female workers. With the intensification of work came calls for relief. Lowell's female mill hands joined a larger movement of the mid-1840s to demand the ten-hour workday in industry. Lowell would become the focal point of a campaign that would grip all the mill centers of New England.

Leaders of the 1840s' ten-hour movement opted for a political strategy. Rather than appeal to individual employers for reductions in the hours of work from twelve and more a day to ten, they sought legislative action, laws that would uniformly impose shorter hours. Organized groups of mill hands launched drives to gain signatures on petitions that were submitted to state assemblies. Activists in Lowell established the Lowell Female Labor Reform Association in 1845, by far the largest of its kind. Within a year's time, the association had assembled a petition addressed to Massachusetts state legislators on behalf of the ten-hour day that included more than five thousand signatures.

The Lowell Female Labor Reform Association mobilized the community of female mill hands in the city for a number of years. The group sent representatives to other mill towns to boost protest activity throughout the region. Association leaders prepared and presented testimony to legislative committees and helped defeat candidates opposed to ten-hour legislation. They held parades and demonstrations and launched a lecture series for mill operatives that featured such well-known reformers and critics as the abolitionist William Lloyd Garrison and the editor Horace Greeley. Finally, between 1845 and 1848, members of the association published the *Voice of Industry*, a popular weekly labor journal that provided women mill workers the opportunity to express themselves publicly.

Ten-hour reformers did not see their demands met. In fact, events were transpiring in Lowell that would see an end to more than a decade of protest by female mill workers in the city. The grueling conditions of work within the mills made employment there less attractive for young women from the New England countryside, and many sought better opportunities elsewhere. Employers who were now finding these Yankee "daughters of freemen" a thorn in their sides and finding their own paternalistic programs too costly and ineffective gradually looked to hire a more docile workforce on a pure and even cheaper wage-labor basis. They turned eventually to the mass of Irish immigrants flooding into

Boston. The demography of the mills and the community soon changed completely; the mill workforce became dramatically more male and foreign-born, and the city came to be composed of separately residing immigrant families. These new social arrangements hardly brought peace to the community, though there would be a long hiatus before the next outburst of organized labor activity. In the meantime, women workers in Lowell provided the first chapters in the history of protest of American industrial workers.

In Lowell, women dominated the labor force of the textile mills and early labor protest. Male workers, who largely occupied skilled positions in the mills, also formed protective associations during the period. Cooperation generally marked relations between the men and the women during periods of protest, especially during the campaign for the ten-hour workday. Lynn, Massachusetts, the other great site of eventual full-scale industrialization in the United States before the Civil War, also witnessed notable agitation on the part of workers; but here the story was complicated by the city's very complex industrial history as well as by divisions within the community between working men and women.

As women in Lynn began to be hired directly on an outwork basis as shoebinders and drawn into wage labor—as they began to labor not entirely as unpaid members of the family economic system, in other words—they picked up the gauntlet of organization. In 1831, women in Lynn and surrounding towns formed the Society of Shoebinders to demand uniform wage schedules for the various upper-parts of shoes they stitched and to encourage cooperation and the resisting of individual bargaining by outworkers. The society boasted a membership of more than two hundred shoebinders. Although the group appears to have had a short-lived history, even the brief organization of women who worked at home across a several-county area represented a formidable achievement.

Two years later, an even larger collective venture was launched. In 1833, in response to wage cuts, more than a thousand women from in and around Lynn joined the new Female Society of Lynn and Vicinity for the Promotion of Female Industry. In a series of meetings and letters to the press, the group made their presence known and felt. Leaders of the movement demanded wage increases and employed the labor theory of value as justification. "It is highly reasonable, that those who perform a considerable portion of the labor, should receive a considerable portion of the recompense," they argued. "Manufacturers frequently become wealthy, while the more laboring portion of the community, are obliged to struggle hard for a compensation, and are frequently distressed." Moreover, women had to be treated with the same dignity as men. "Equal rights should be extended to all—to the weaker sex as well as the stronger."

Lynn's shoebinders echoed the language of protesting craftsmen to a greater extent than their sisters in Lowell. For Lowell mill operatives and machine tenders, anger over their poor treatment flowed from their perceived position as daughters of independent farmers and freeholders. The manufacture of shoes still remained in the hands and fine stitches of Lynn's shoebinders, and grievance emanated from the pride they took in their work. The women shoe workers of Lynn, of course, were not like craftsmen in one great respect. Opportunities for masterhood were closed to them. When the women demanded just wages, they sought "comfortable support" for their families, "freedom from want," and in the case of widows, avoidance of dependency and public welfare; it was compensation to bolster the family, not the male demand for a "competence" that would allow for eventual independent producership.

Members of the Female Society of Lynn received assistance from men shoemakers, who had formed their own Society of Cordwainers in the early 1830s. That group agreed to refuse orders for lasting the uppers to the shoe bottoms they prepared from manufacturers who had not acceded to the wage demands of the shoebinders. Despite that support and a flurry of protest activity, Lynn's women shoe workers faced an almost impossible task in organizing their trade. Manufacturers could simply transport materials to be sewn through a widening arc of the New England countryside to households that needed the extra work to make ends meet or that wished to purchase goods in the region's expanding market. The female society collapsed within a year's time.

Relations between men and women workers in the shoe industry became more complicated in the decades that followed. In the 1840s, male shoe lasters reorganized into an effective cordwainers' union that published a popular labor newspaper, the *Awl*. The men invited the female shoebinders to join the association but on an auxiliary basis—to lend support and have the organization appear as a defender of working families in the community, not necessarily to help them exert their own pressure on employers. The question of whether women were to act and be treated as independent wage earners or as members of family economic units that were led by men emerged as an issue in a full-blown and telling way in labor disputes in the 1850s, including the shoemakers' general strike of 1860, the largest strike of American workers in the antebellum period.

Changes in the organization of production in the shoe industry provide the backdrop to events in the 1850s. Increased centralization and mechanization during the decade led to the creation of an industrial shoe labor force to complement the shoebinders and shoemakers who continued to work on a contract and outwork basis. Industrialization writ large with migration to the community drove down the price of labor—and an economic depression in 1857 made matters worse—but the responses among workers varied significantly. In 1859 a protest movement led by journeymen shoemakers emerged in Lynn and other shoe towns. The men set forth their demands for wage increases. Female factory shoebinders then placed their own agenda on the table, and an entangled debate ensued.

The factory women were largely young, unmarried, and new to the community—without roots in the old home/shop system of production. The shoemakers asked the factory women to lower their demands; they feared that if factory wages were raised, stitching of uppers would be centralized and homework eliminated, thus jeopardizing family incomes. They argued that efforts should instead focus on increasing men's income. The factory women, under the leadership of the twenty-one-year-old Clara Brown, countered by declaring their willingness to organize and represent outworkers, a show of female solidarity. Community pressure mounted for the factory women to lower their sights, however, and eventually a majority voted to follow the men's leadership. An official strike was then called to begin on Washington's Birthday in late February 1860.

By early March more than ten thousand shoe workers across New England joined the cause, with Lynn providing the anchor for the movement, and on March 7 Lynn was the site of a grand protest parade that became the largest labor demonstration of the early industrial period. Women shoebinders proudly marched under the banner, "American ladies will not be slaves. Give us fair compensation and we will labor cheerfully." The great show of solidarity hid great divisions—and an actual lack of strength, for the strike would crumble within a month's time with only scattered gains. The shoemakers' strike of 1860 revealed shared grievances and commitments among workers, but also fractures within the working class—between workers tied to and protective of old ways and their neighbors already enmeshed in the wage-labor, factory production system. Wrapped in this broad divide were the cutting forces of gender and ethnic differences. Industrialization wrought camaraderie and protest within working-class communities and serious breaches as well.

Other Responses to Industrialization

The labor organizing and protest activities of craftsmen and industrial workers—women notably figuring in the latter instance—during the antebellum period represent the overt responses to industrialization. Americans also reacted in less defined and more personal ways to the great economic and social transformations of the day. The expansion of market activity, the spread of the wage labor system, mechanization, and the coming of the factory disrupted community life. While people responded to the general upheavals of the times by mobilizing in certain instances to create new institutions of social order, they also responded personally as well by affecting new manners and appearances. As with the more explicit responses of workers to changing conditions in the workplace, machinery and technology were not the fundamental issues of concern; rather, changes in relationships among people within communities forced these other kinds of reactions.

Transiency was one visible sign of changing times. People moved within and through American antebellum cities and towns with great frequency and in great numbers. This movement has to be placed in a world-historical perspective. The nineteenth century was marked, at least in the Western world, by massive labor migration flows. The passage of laws after various democratic revolutions allowing men and women to migrate from their homelands, and concurrent advances in transportation and communication facilities, *enabled* people to move. Religious and political persecution, deteriorating economic circumstances, but most important, changes in agriculture and rural landholding patterns *pushed* families away from their places of birth. New opportunities in industry and expanding urban economies in general finally *pulled* people from age-old areas of settlement. The United States was the farthest point west in this migration swirl, but the process continued on the American mainland. With the country's expanding frontier, some immigrants—especially Scandinavians and Germans—found themselves back on the land. Others were lured to urban centers, as they had been in Europe. As the century wore on, the country witnessed further westward movement but also a rural-urban migration flow that mirrored developments in Europe.

Scholars have fully documented the notable transiency of the American people in the antebellum period and throughout the century. The extent of movement and the groups that moved the most have been calculated and determined. Yet, the reasons for migration and the implications for the lives of those involved remain clouded. Historians have thus selected samples of Americans from U.S. manuscript population censuses and noted who could and could not be found ten years later in the next official censuses. Upwards of 60 percent of traced populations invariably disappear from the records, an indication that only two in five Americans stayed in their communities for extended periods of time. These figures hide the actual extent of population turnover, however, for not taken into account in these investigations are the sizable numbers of people who moved in and out communities within the intervals between census enumerations. Another kind of tracing conducted by historians can render this greater sense of the churning of the American population. In following workers' careers through early- and late-nineteenth-century payroll records of companies, scholars have discovered in any number of instances that more than 50 percent of those employed stayed with their employers for periods of no longer than six months. Whether through census or payroll tracings or available literary evidence (letters, sermons, commentaries of the day) all indications point to a population constantly on the move, to high levels of transiency.

We know who moved most regularly—young, single adults and people without property—but why these or other Americans chose to migrate and with what consequences are puzzles not easily solved. Did Americans move in such great numbers because of poor prospects and failure? Or to ensure ever better opportunities? Did they find securer circumstances? What impact, if any, did migration have on them and their families? Since tracing people across locations is a difficult task, historians have not been able to answers these questions with any certainty. It is known that moving represented for some a way of "voting with one's feet"; workers left incorporated mill villages in New England and the Middle Atlantic states, for example, to escape the regimen and close supervision. This was an unorganized and personal form of labor protest.

Transiency appeared as one mark of the unraveling of the social order; greater differentiation among the American people loomed as another. Before the nineteenth century, American communities were hardly homogenous. The ethnic and racial mix of the population was already pronounced, and dramatic inequalities in personal and real wealth emerged and prevailed at early dates. Certain rural communities, most notably in New England, may have had a uniform look, but diversity and inequality characterized settlements in North America from the outset.

Differences became more glaring during the early industrial period, but that is only a part of the issue. Pre-nineteenth-century American communities may have been heterogeneous, but they remained integrated with people from all backgrounds and walks of life comingling in close proximity. By the mid-nineteenth century, eyewitnesses could not be but impressed by the segmentation within American communities; a districting of difference had occurred that disturbed older and established members who feared for the decomposition of their cities and towns. Population expansion and dispersion and immigration guaranteed a greater balkanization of society; communities were now too large for face-to-face relations among different groups, and the very possibilities for ghettoization had increased. Also critical was the role that the spread of

market activity, the wage labor system, mechanization, and factory production—industrialization writ large—played in creating grave disparities in position and means.

In the new, diversified manufacturing centers of antebellum America,
the segmentation of social life took its most visible form. Population expansion there forced residential dispersion and real estate development.
The spreading outward of city dwellers was not a random process; income,
in fact, became the key determinant in where an individual or family
would land on the expanding urban map.

The poorest of the inhabitants of New York, Philadelphia, and elsewhere clung to neighborhoods about the docks and in downtown in
general. Employed as common day laborers and on an intermittent basis,
they stayed near to the greatest concentration of unskilled jobs, walking
about day after day if need be in search of work. As city dwellers with the
least means, they also could not afford the costs of commuting on the
new, privately owned horse-drawn omnibus systems that were laid out
in American urban centers during the antebellum period: the daily fare
would have amounted to one-tenth of the seventy or eighty cents a day a
common day laborer then earned. Lower-class neighborhoods thus had
emerged in all American cities by mid-nineteenth century, inhabited largely
by young single males, newcomers, both foreign and native-born, who
lived in crowded boardinghouses and tenements.

Away from the teeming dock areas there appeared factory districts.
Manufacturers, whose operations outgrew their downtown central shop
spaces, looked toward nearby outlying areas to build larger facilities. Workers who were more steadily employed and their families began to inhabit
the neighborhoods that developed around the mills. They lived in rented
apartments and houses built by either their employers or small-scale real
estate operators, or in small row homes that they purchased with loans
from local savings societies or fraternal organizations. The new mill districts formed enclosed, relatively settled communities with local churches,
schools, retail businesses, and ethnic and labor organizations.

Outward from the port and industrial districts, other kinds of residential neighborhoods emerged. Skilled workers who could afford transportation and the purchase of modest row or detached homes moved to new
respectable working-class communities. Professionals and small proprietors bought larger homes on larger lots of land farther away, and at
the borders of the city by the 1830s and 1840s could be found the first
suburbs of large estates built by merchants, bankers, new industrialists,
and other people of means. The wealthy could not only afford the commuting costs, but they could also take advantage of decreasing land values at the farthest points from the commercial downtown to construct

their suburban mansions on vast acreages. Their move outward also represented an escape from the growing unpleasantness of city life.

By the mid-nineteenth century, social divisions within American communities could be mapped, every ring about the center representing a different income group; even in districts where a certain heterogeneity prevailed, internal distinctions occurred, with the wealthy living on the avenues and people of lesser means on the side streets. Segmentation appeared in its clearest form within cities, but these developments were matched in smaller communities as well.

The separation of people within localities had a number of effects. As one example, in the settling process within communities, work became divorced from home life. A good portion of the community now lived at significant distances from their work. This changed the place and meaning of work (and leisure) in their lives and led to a system of dual identities: people were now members of separate work and residential communities. This would have great implications for labor protest, as many scholars have argued, and for politics in general.

The new spatial and sociodemographic arrangements of communities, in fact, had great impact on politics. By the third and fourth decades of the nineteenth century, cities and towns in the United States had become too large and populous to remain administered through town meetings. Accordingly, representative systems of government were adopted. As families of wealth and standing abandoned the city for its outskirts, vacuums in power also emerged within antebellum communities. From newly created political districts, ambitious local leaders appeared to fill the void. They characteristically built neighborhood organizations to get out the vote; once in power, they returned favors to their political allies and sought and provided a variety of services to their local constituencies. Competitive ward politics thus replaced the deferential politics of an earlier era when communities were compact and people bowed to well-recognized elites.

This new system of politics quintessentially materialized by mid-nineteenth century in New York City with the rise of the famed Tammany Hall machine of the Democratic Party and was met with both apprehension and disdain by people of standing. The street-based, street-wise politicians of the party also vied with labor activists; at stake were the attention and allegiances of working people. But the seemingly sympathetic politics of the men of the political machines were a far cry from the politics articulated by the men of the shop, and not just because Democratic Party ward leaders often formed alliances with business elites in the community; rather, their rough-and-tumble, pluralistic politics, a politics to match the highly self-seeking and competitive economic order then emerging, struck at the heart of an ideal, that of a republic of independent, virtuous citizens and producers.

Disparities in income drove the segmentation process, with its various social and political consequences. Even with population pressure and transportation innovations, the placement of people within communities would have been more random and less pocketed had wealth been divided equally; Thomas Jefferson and other spokesmen for the republican vision could have warned as much. The issue of income also raises another question regarding standards of living during the early industrial period. In Great Britain, generations of scholars have debated whether industrialization led to the pauperization and immiseration of substantial portions of the British population. No consensus has been reached on the subject despite substantial research. Students of nineteenth-century American social and economic history have not engaged in great discussion on the matter (particularly for the antebellum era). Research is limited, and any comments on the material well-being of the first generations of American wage laborers have to be tentative.

Numerous methodological and conceptual problems confront investigations of standards of living during the antebellum period. Only scattered records of wages and commodity prices survive for the era; wage information, moreover, reveals little about total incomes, since a significant portion of the workforce was employed on an irregular basis. (Government agents would not collect data on wealth and income in a systematic fashion until after mid-century.) Enormous variations in wages across and within regions, across occupations, and over time also have to be considered. Conditions were particularly dire in the late 1830s and early 1840s, when the country was in the midst of a severe depression. Finally, calculating the base needs of a typical family of four in order to determine the percentages of families living above and below subsistence levels raises difficult interpretative questions.

The following has been established on the subject. American workers during the antebellum period earned substantially higher wages than their counterparts in Europe; skilled workers, in particular, were paid at rates two and three times higher than they would have been in England, Germany, or France. The flow of people west across the Atlantic is testimony to the differential. American workers also benefitted from the rich, expanding agricultural base of the country; food prices were substantially lower on this side of the Atlantic. Immigrants reported in their diaries and letters home of having meat meals two and three times a week—absolutely unheard of in Europe—and of being able to afford wheat rather than oaten bread. "You would be surprised to see provisions so cheap," John Parks, a carpenter wrote to friends in England in 1827;

"we buy the best of meat for 4p. per pound." John Mooney, another immigrant, echoed his words, saying that the "food of the American farmer, mechanic, and labourer, is the best, I believe enjoyed by any similar classes in the whole world." American workers also benefitted from the cheap prices of other goods, particularly cloth that issued from the fully integrated and mechanized mills of New England.

Evidence of relatively high wages and low commodity prices can lead to the conclusion that American workers in the early nineteenth century had the good fortune in general of living above the margins, but definite qualifications are in order. First, when wages and prices are charted in tandem between 1820 and 1860, only modest gains in real income are recorded; this was not, then, a period of rising standards of living. Second, wages varied dramatically. Unskilled workers earned between sixty and eighty cents a day during the era; women textiles operatives earned at the low end of this scale, between fifty and sixty cents a day. Skilled workers, in contrast, could expect to receive at least two and three times the average daily wage paid common hands. Some aristocrats of labor at the time—iron puddlers, for example—could even make nearly three dollars a day. These variations obviously militate against drawing any general conclusions concerning the material well-being of American people during the antebellum period. Finally, American workers may on average have received greater compensation and benefitted from lower food prices to allow them to live at higher standards than laboring men and women in Europe, but this may have also come at the cost of longer hours of work and an intensified pace of work. Visitors to the United States who came to see the dazzling aspects of American industrialization, as well as new immigrant workers, remarked consistently on the different regimen of the American workplace. "I can assure you I never worked so hard," William Danley wrote back to his wife in England in 1857. A Norwegian tin worker warned his relatives, "Day labor [in New York City] demands a more strenuous exertion than we are use to." A reporter for a British journal placed the matter squarely: "Although [the emigrant] may get better wages, he has to give a much greater amount of labour for his money." Statistical studies have not been conducted to determine whether higher standards of living in the United States were in fact related to intensified work, but the literary record is clear on the question. It is not coincidental that hours of work represented such a key and constant grievance of the early trade union movement in America. In the labor protests of the 1820s and 1830s, wages were rarely at issue; hours and other concerns about the place of working people in the republic were uppermost in mind.

The "standard-of-living controversy" is a basic component of studies

on European industrialization during the nineteenth century; the subject
has not found a place in American histories. This reflects the common
judgment that by and large, the first generations of American wage earners lived above basic subsistence levels and with some comforts (even
if that required extraordinary labors). However, that general conclusion
has to share the spotlight with another: that extreme disparities in wealth
and income prevailed in the period. Unskilled workers who earned eighty
cents a day—even if they worked a full complement of days over the year,
which was highly unusual—could not have fed, clothed, and sheltered
their families, except in the barest of ways, on their $250 yearly incomes.
Commentators at the time estimated that families of four needed at least
$400 a year to live with a modicum of decency; some placed the figure as
high as $600. The only way for these families to survive was to have two
and three extra breadwinners in the labor market, which meant the
employment of older sons and daughters (and in fact, despite the growth
of public school facilities, the nineteenth century would witness progressive increases in teenage participation in the labor force, particularly
those teenagers from families whose fathers occupied jobs on the lower
end of the occupational scale). Nor did many unskilled workers succeed
in improving their lots during their lifetimes. Scholars who have calculated rates of occupational mobility have found that when Americans in
the nineteenth century moved up the occupational ladder, they did so
only in small steps.

Inequalities had marked American settlements from the outset, but the
situation by the mid-nineteenth century had new dimensions. All evidence points to a widening of difference. Wage rates between the skilled
and the unskilled spread during the antebellum period, as did the gap
between salaried workers, professionals, and employers in the community and wage earners in general. The expansion of market activity,
the extension of the wage labor system, mechanization, and the building of factories had created a growing tier of new, low-paid positions,
jobs often filled by women, children, and immigrants. The bottom had
swelled, and people at the bottom now formed a visible mass in the core
of American cities and many other communities. It was the glaringness
of the inequalities, not the discrepancies themselves, that was new.

Americans separated and clustered themselves within their communities during the antebellum period. They formed enclaves defined by
specific levels of income. But the divisions that emerged in the population
at the time were not purely economic. The splits between the various
haves and have-nots were also divides of culture, of habits, and of manners. Differences among Americans during the early industrial era can be
traced to differences in work experience and personal means, but varia-

tions in urban-rural and ethnic backgrounds figured as well. This is to say that there would have been disharmonies within American communities because of growing class divides, but the fractures were compounded by related divisions in cultural heritage and traditions.

Within working-class wards alone in antebellum cities and towns, visitors could discern at least three distinct subcultures. Men of the craft shops maintained a world of their own. They proudly marched together in civic celebrations displaying their wares and in protest parades. They convened in meetings and lectures to hear about and discuss the latest developments in science as well as new social and political ideas—for example, on the latter, producers' or workers' cooperatives, utopian socialist communities, and land distribution reforms. They tried to salvage the remains of their trade union organizations, which were decimated by the economic crises of the late 1830s; and by the late 1840s and early 1850s they presided over a revival in labor activism that accompanied a return of prosperous times. They were joined in this new surge of organizing by immigrant skilled workers from England and Germany, many of whom had participated in radical movements in Europe and brought with them new energies and commitments. The activist craftsmen formed a distinct segment within working-class communities; they were serious, literate, secular, highly politicized men who played a role disproportionate to their numbers in mobilizing their neighbors and providing leadership, ideas, and rhetoric for organized labor protest.

Another segment of American wage earners during the early industrial period found religion. Religious revivalism spread in wildfire fashion through American communities during the 1830s and 1840s. Those who were inspired established organizations that blanketed the nation with Bibles and religious tracts; makeshift churches appeared overnight, and a score of fiery ministers tramped from town to town bringing the word of personal redemption through faith and renunciation of sin. There is no easy way to account for the sweep of religious enthusiasm during the period. At a time of great social and economic transformation (and economic crisis, with the eight-year depression starting in 1837), some Americans may have found comfort in intense religious experience. For older-stock Protestant Americans, fervent belief may have brought a semblance of order to their lives, especially in the face of industrialization and the immigration of massive numbers of Irish Catholics. In challenging staid established churches, the new revivalist ministries also mirrored the new open and competitive spirit of the times. Explanations of this kind are available, but no sociological answer to the question of evangelicism can do justice to the engagement of any single person.

Religious revivalism of the antebellum period is commonly associated with rural and small-town America and people of middling social rank, yet Protestant workers in industrial areas enrolled in the mission. Recent research suggests that evangelical ministers found a ready audience in mill towns and urban factory districts. The audience may have consisted of Protestant men and women who had just migrated from farm areas and found fervent religion helpful in coping with industrial work and regimen; it is also clear that in many instances, their attendance at church and religious meetings was required by their employers. Only a mixed assessment as to the backgrounds and motivations of engaged workers is possible. Nonetheless, the historical record does show that working people did attend revival meetings, join new churches, declare their piety, and, most notably, publicly forswear drink. The evangelism of the period gave rise to a vocal and visible temperance crusade, and Protestant workers swelled the ranks. The Washington Temperance Society, founded in 1840 by a group of artisans, proselytized directly in working-class neighborhoods; and organizers boasted that more than three million workers had sought membership in the association by the mid-1840s and in the process declared their abstinence from alcohol.

In seeking to lead pious and sober lives, working-class evangelicals marked themselves off from labor activists, but also from a third contingent within the antebellum laboring population. A more raucous mass of working people lived at the margins of Americans communities at the time. Generally engaged in heavy physical labor or sweatshop work, employed irregularly, newcomers to the industrial scene, newcomers to America in many instances as well, they acted in freewheeling and episodic ways. They worked hard, but set the time, pace, and standards of their work. They also played hard. By the mid-nineteenth century, American communities had their rough-and-tumble wards known for their saloons, amusement halls, unruly fire brigades, street gangs, and rich street life (the Bowery in New York City became the most notorious). The character of these neighborhoods can be explained in part by the kinds of work available to their inhabitants, in part by their age profiles—young, unattached men and women composed a greater part of these subcommunities—and in part by their ethnic composition: non-Protestant immigrants of rural background increasingly populated these districts, and they generally held to task-oriented rather than clock-oriented senses of time. The growing visibility of this community was a function of their numbers: while the foreign-born represented 20 percent of the populations of such cities as New York in 1820, they were 50 percent of the population of New York by 1850; 400,000 immigrants came ashore in the city in the decade of the 1840s alone, and they would constitute upwards of 80 percent of the wage-earning workforce.

Complicated, often tense relations prevailed among the subcommunities of the laboring population. Men of the shop and the evangelicals shared a common sober disposition toward life, but not a politics. A number of union leaders could trace their involvement in labor protest to intense religious experiences; the democratic, antiestablishment, and perfectionist impulses of evangelism could spur militancy on behalf of the oppressed. But for most religious workers, the search for personal redemption led away from politics and toward conservative attitudes and behaviors.

Relations between the pious workers and the improvident were completely hostile. The divides between these two communities were manifold. The common hands of the community were seen as threats to the wages, jobs, neighborhoods, and sensibilities of those who sought respectability. To make matters worse, age-old antagonisms between English Protestants and Irish Catholics were involved. In the early 1840s, amid worsening economic conditions, the situation turned nasty, with Protestant workers joining a series of nativist attacks on immigrant Catholics in various cities in the Northeast. Violence also erupted from within lower-working-class neighborhoods, as the same period witnessed rioting between immigrant groups and African-American day laborers who occupied the same poor and crowded areas of communities and vied for the same menial jobs.

Between labor activists and the downtrodden ambivalence marked relations. Unskilled workers represented a forceful mass whose anger could be tapped on behalf of organized protest. During the general strike of 1835 in Philadelphia, and at other moments, labor leaders reached out and mobilized this portion of the laboring community with notable success. Yet here, too, there were enormous divides of occupational experience, political perspective, and temperament. Local mainstream politicians were better able to forge allegiances with workers from lower-class neighborhoods, especially immigrants, and attach them to the machinery of the Democratic Party—thereby blunting independent labor party efforts.

The Emergence of a Middle Class

The divisions manifest among laboring people in American communities during the antebellum period had repercussions far beyond working-class neighborhoods. The divides among workers—based on varying work experiences and incomes, conflated by cultural differences, and reinforced by geographical segmentation—handicapped trade union organizing and labor political activity. For community members of greater wealth and standing—shopkeepers, small manufacturers, new white collar workers, professionals, merchants, industrialists, and bankers—the splits within the laboring population were signs of the general breakdown of the social order. Whole sections of their cities and towns had emerged as alien and beyond control. Rootless young men and women, disheveled and unlikable immigrants, and angry mobilizing workers comprised these districts; the crime, low life, and internal violence of these areas threatened to spill over into respectable neighborhoods; and with the system of ignoble ward politics, political leaders could no longer be relied on to tame the unruly. The unraveling of community relations accompanying the great economic and social transformations of the day produced a response among people of means. Middle-class members especially reacted in formal ways: they organized to create a wide array of associations and institutions of social order. They also responded in personal ways: they moved away from alien neighborhoods to their own enclaves, found comfort in religion, and adopted new styles of behavior to mark themselves apart.

The early industrial age, from the 1820s to the Civil War, witnessed a mobilization on the part of people of means within American communities. Foreign visitors to these shores at the time (the French historian and social commentator Alexis de Tocqueville stands out here) made special note of the unusual proclivity of Americans to form voluntary associations; they referred by and large not to the trade union organizations established by workers, but to plethora of organizations created by Americans of middle- and upper-class standing. During the antebellum period, these Americans participated in a flurry of associational activity.

Religious organizations topped the list. In community after community during the antebellum period, well-off citizens (that is, Protestants) formed local Bible societies, Sunday School unions, Sabbath observance groups, religious tract societies, missionary associations, and, without fail, temperance organizations. Spiritual reawakening obviously required more than the spirit. Religious fervency also led to other kinds of associational work. Religiously inspired Americans helped form various private charity groups and lent their time and money to such causes as the abolition of slavery, greater rights for women, the creation of free common schools, and prison and asylum reform. For middle- and upper-class Americans this was an era of association as well as institution-building, and they contributed to the establishment of school, police, prison, and mental hospital systems in this country.

People of means participated in other kinds of associational activity. They formed literary, drama, and music societies; the upper crust established their exclusive sporting and social clubs. They also helped to establish new political movements in the country. Some became involved in nativist politics, and through both organizations and political parties, called for severe restrictions on immigration. Others participated in the building of first the Whig Party and, later in the 1850s, its successor, the Republican Party. Both parties stood for a dynamic and organized society, encouragement of industry, greater state intervention in economic and social life (for example, government-mandated tariffs and temperance), and rule by the meritorious. Whigs and Republicans presented another postmercantilist possibility, and they appealed to religiously inspired Protestant workers on economic and moral grounds to forsake labor radicals and Jacksonian Democrats alike. Tariffs would protect their jobs, they would oppose the extension of slavery (and the rule of aristocratic slaveholders) in order to keep western territory open for yeoman producers, and they would bring order to communities through their example of rectitude and policing legislation.

The association- and institution-building activities of middle- and upper-class Americans during the antebellum period can be simply labeled as efforts at social control. That conclusion is unassailable, yet a great deal of history and understanding is lost if some perspective and qualification is not added.

American communities had become too large, diverse, and segmented for order to be instilled by consensus reached at town or church meetings or through deference to known elites. Those interested in reordering their communities had to create new institutions, and that required organization. That much is clear. Yet, religious revivalism added a dimension and complexity to the process. Well-off Americans entered a period of institution-building at the exact moment of a spiritual reawakening. Whether the latter was more cause than effect can occupy hours of unresolvable debate, but the point is that this particular effort at social reordering was colored by particular religious ideals and zeal. Antebellum Protestant evangelism had a positive, perfectionist core (it shared much with Enlightenment thought), and religiously inspired reformers sought to rehabilitate and not simply to discipline and dominate. They established associations and institutions with reformative ends in mind (the later purely custodial nature of the institutions formed had more to do with the bureaucrats placed in charge than with the intentions of the originators). Religiously inspired reformers also found themselves engaged in activities that aimed at greater social equity rather than just order. Their involvement in abolitionism and women's rights causes brought disfavor and attack and was hardly motivated by the longing to quell social tensions. Finally, a focus on middle- and upper-class reformers loses sight of support for reform from other groups in the community as well as opposition to their efforts. People of property may have petitioned for the establishment of public school systems to ensure a future workforce of disciplined wage earners, but organized workers lobbied for public schools too for their own reasons. There were limits, as well, to the impact of the reformers. Try as they might, religiously inspired temperance crusaders could not curb drinking in working-class communities. There were walls of culture and geography that they could not breach.

In engaging in associational activity during the antebellum period, the well-off of American communities may in fact have had greater impact on their own lives than on those whom they attempted to alter. The organizing of middle- and upper-class community members should be understood as an effort at self-definition as well as social control. In appearing as the purveyors of right behavior and sustainers of the community, they created particular personas. People of means reacted to the great economic and social transformations of the early industrial age by distinguishing themselves. They did this through religious and secular organizational endeavor, and this is another way of understanding their activity.

Americans of property and standing also consciously marked themselves in more private ways and in particular by creating a new kind of home life with an emphasis on propriety and respectability. Here, two developments are of importance: the growth of middle-class occupations and the changing role of women. The expansion of market activity enlarged the ranks of people of middling status. Retail shopkeepers, small manufacturers, professionals, and a growing army of clerks and other white-collar employees now comprised a identifiable population in American communities, and it would be they who would labor to fashion lifestyles of decorum. They modeled themselves not in a vacuum, but in opposition to those above and below them—to both the profligate, indulgent rich and the improvident poor.

Mothers occupied a key place in the new respectable middle-class household. (And this was a smaller household, it should be noted. As a measure of the new emphasis placed on social and personal control, middle-class families led the way in a general demographic shift in the mid-nineteenth century that would see—before widespread knowledge and use of contraceptives—the lowering of birth rates of women from more than five children, on average, to slightly more than three.) Men now worked outside the new ideal small household; their destinies lay in the marketplace and in earning sufficient incomes to maintain their families in comfort. Women's place was in the home. Mothers remained responsible for the traditional hard work of the home—cooking, cleaning, sewing, and in some instances in middle-class households, the taking in of piecework—but the popular literature of the day also prescribed for them the role of guardians of the family, in charge of the moral education of children and the upholding of family standing. As to the latter, women oversaw the tasteful furnishing of their homes, once sites of production and now of consumption (though not too conspicuous). Men and women were thus to play separate roles. As one antebellum advice writer counselled, "Each has a distinct sphere of duty—the husband to go out into the world—the wife to superintend the domestic affairs of the household."

An ideology thus emerged during the early nineteenth century that limited women to the private world of the family. Yet, as with plans for the Yankee farm girls who went to work in the textile mills of Lowell and elsewhere, women defied the conventions. A separate sphere presented women of the middle class with opportunities to form close friendships among themselves—they refused to remain confined to the home—and also with a great deal of time to become engaged in religious and reform work. In fact, the history of religious revivalism and associational activity in general during the antebellum period is largely a history of women, for women constituted a disproportionately large share of the membership and often leadership of various movements. Women were denied citizenship and relegated to the domestic sphere, yet in actuality they played a most public role in their communities.

The respectable middle-class home was another response to and product of the early industrial period. People of means during the antebellum period mobilized to create organizations and institutions that aimed at restoring order. They forged lifestyles that brought some stability to themselves and their families in the face of growing social divisions and turmoil and the emerging competitive spirit of the age. A catalogue of reactions to industrial development includes middle-class as well as labor activism.

A small group of Americans responded in a different way. They reacted to the coming of unbridled market activity and the factory system by opting to live in utopian socialist communities, places where all property was owned in common, the work shared, and the proceeds of the work distributed equally. In some instances, as in the case of the Shakers, these communities were established by religious groups; others, such as Robert Owen's New Harmony, founded in Indiana in 1825, were entirely secular ventures. Few of these experiments in communal living survived, yet by the time of the Civil War, no fewer than 150 utopian communities had been formed in the United States.

The adoption of a communal lifestyle represented perhaps the most radical of the responses to the economic and social transformations that altered America in the first half of the nineteenth century—to the changes that are conveniently labeled "industrialization" or "early industrialization." The history of the period reveals a general apprehension about the expansion of market activity, the spread of wage labor, and the factory system. The machine was not the issue. Americans greeted the new technologies of the day with curiosity and enthusiasm: the machines could serve, not displace. The changing nature of social relations and the breakdown of community life caused alarm. Americans of different standing reacted in different ways, both formally and informally. But the reactions were of a tentative, private, or local kind. New organizations, many ephemeral, such as trade unions and religious and charity groups, appeared, as did new institutions—the middle-class family and public schools, for example. The overall forces of industrialization were thus mediated but not stemmed. Events had unfolded in too uneven a fashion for there to have been a general or definitive response. But more important, in dismantling an old order of privilege and control, Americans of varying interests collectively ushered in the new competitive age. They did so warily, but antimercantilist sentiment prevented—at least for the time being—a more structured response to the upheavals of the day, especially a response that would have involved the powers of government and particularly their national government.

Chapter Four

The Civil War and the
Politics of Industrialization

THE HISTORY of American industrialization can be rendered with but fleeting reference to the American Civil War, that cataclysmic and central event of the American nineteenth century. The expansion of market activity, the spread of wage labor, mechanization, the coming of the factory system, the massive migration of people into and through the country, urbanization, occupational change, social segmentation and divisions within communities, labor organization, the emergence of a distinct middle class, the separations between men and women, declines in fertility, and ultimately, the rise of the large-scale corporation—all of these fundamental nineteenth-century socioeconomic and demographic developments in the United States can be described and analyzed without mention of the Civil War. And, in fact, American social historians writing in recent decades have, with few exceptions, bypassed the war in their histories of American social life; this, despite the full knowledge that the war freed almost four million people of African-American heritage from bondage, resulted in the deaths of more than 600,000 Union and Confederate soldiers, and left no community in the United States untouched by its physical and personal effects.

For earlier generations of American historians, the Civil War occupied a prime place in the nation's history and especially with regard to industrialization; those historians would find curious the tendency of present-day historians to downplay the war's importance in broad social transformations. In one classic rendition of the country's past, the war and industrial development are, in fact, pictured as inextricably linked. This view can be outlined as follows. Northern manufacturers early in the nineteenth century attempted to construct an industrial base for the country, but they faced hostile political circumstances. To succeed, they required assistance from the federal government in the form of protective tariffs, a national banking system, state-supported transportation improvements, and other public encouragements. Southern politicians, representing the interests of the planter elites of the South, however, repeatedly blocked all activities of the federal government on behalf of northern industry; they objected both to the costs involved and the infringements on states' rights. The vying economic imperatives of the two regions, the need to control the federal government, and related differences over the question of the extension of slavery made pitched conflict between the North and the South inevitable. Once the Civil War began and southern representatives withdrew from the halls of Congress, northern politicians could and did write a legislative program that fostered and guaranteed industrial development. The Civil War, in this way of thinking, represented the "triumph" of industrial capitalism, a necessary matching of the political system with the emergent industrial economic order.

The place of the Civil War in American economic history will be assessed in this chapter, but only as a means toward a greater end. The issue of the war and industrialization provides an opportunity to address the general question of government and industrial development. What role, if any, did the government play in American economic growth and development during the nineteenth century? Was state intervention essential, or did industrialization have an inherent life and force apart from politics? And specifically as to the Civil War, did the war matter in the grand economic and social transformations of the day? An investigation of the politics of industrialization will require a look at specific government initiatives as well as an examination of the indirect impact that the nation's political and legal structures had on the economic life of the country.

Government and Transportation Improvements

Tench Coxe led the first charge on behalf of American manufacturing. In the 1820s and 1830s a second generation of advocates emerged to boost the cause of industrialization, but their message had a different twist. They explicitly called for direct government assistance to the nation's new industries. The chief spokesman was Henry Clay, who over a long political career represented the state of Kentucky in both halls of Congress and three times aspired to the presidency of the United States. From his earliest years in public service, Clay championed manufacture. In 1808, as a young member of the Kentucky state legislature, he sponsored legislation requiring state officials to wear only American-spun clothes. In 1810, as a new member of Congress, he argued that the Navy should be prevented from purchasing foreign supplies. He delivered his most important statement, however, in 1824, in a speech to Congress advocating passage of strict tariff measures to eliminate foreign competition and protect American industry. He spoke then of "adopting a genuine American system," and for a generation the phrase "American system" became identified in American politics with calls for government protection of industry and sponsorship of internal transportation improvements. Joining Clay in building sentiment on behalf of government support of industry was Daniel Webster, long-serving senator from the state of Massachusetts, greatest orator of his day, and spokesman for New England manufacturing interests. Webster also became a public champion of the corporate form of business enterprise; corporations, he argued, could marshall the resources to bring national economic supremacy and prosperity to all Americans. Tariffs and internal improvements specifically, and government encouragement and protection of large-scale business ventures generally, became staples in the Whig political party movement that Clay and Webster founded and led during the antebellum period.

The Whig economic program met with significant opposition. Yet, the antebellum period did witness important government initiatives on the economy, especially in transportation improvements. Here, background discussion is in order.

A transportation revolution accompanied America's march toward industrialization. Greater market activity and an open flow of people, goods, information, and money required the removal of various physical obstacles to transport. Peculiar features of American geography made this a keen problem. The country possessed a vast western frontier; as people settled in the West, great distances separated them from eastern (and western) ports. Exacerbating matters was the almost complete absence of natural east-west waterway systems in New England and the Middle Atlantic states as well as mountain ranges that limited interactions with settlers in the Midwest. In these areas there was a need for extensive manmade transportation networks. (In the South, because ships could travel far inland on rivers, transportation improvements loomed as a less pressing issue.)

Americans moved to open the West to settlement and commerce initially through the building of roads. In the 1790s, turnpike construction created thoroughfares across the Mohawk valley in New York, the southern tier of Pennsylvania west from Philadelphia, and through Virginia.

By the first decades of the nineteenth century, whiskey distilled in Tennessee and Kentucky could pass by wagon to eastern cities; herds of cattle similarly were driven east out of Ohio on improved roadways.

Wagons could not possibly handle the volume of goods produced in the West by the third decade of the century; nor could they get products to the market in a timely fashion. Americans next looked to improving river transport—dredging and widening riverways and experimenting with steamboats—but where navigable rivers did not exist, a dramatic initiative was entertained: to cut through the earth and create manmade waterways or canals. A canal-building boom transpired accordingly in the 1820s and 1830s. By the end of that period, 3,326 miles of canal had been constructed. The inspiring Erie Canal built across upstate New York connected the Great Lakes with the Hudson River and the port of New York City. Other canals stretched westward across New Jersey, Pennsylvania, and Maryland; and in the 1840s, building was completed on a remarkable set of canals reaching north and south through Ohio and Indiana linking communities from the Ohio River to Lake Erie. All this construction occurred at great expense and with the great pick-and-shovel labor of armies of Irish immigrants.

Developments did not stop there. As zealously as Americans embarked on canal building, they moved even more fervently to crisscross the country with railroads. Starting with the laying of track in 1828 in Maryland and the beginnings of the Baltimore & Ohio Railroad, the nation would witness several surges of railroad construction. As early as 1840, the United States surpassed Great Britain in track mileage, with 3,000 miles of track constructed at a cost of $75 million; Americans once again had taken technologies developed in Britain and innovated at a faster pace. By 1850, railways stretched across the Appalachian range. In an unprecedented boom, in the next decade more than 20,000 miles of track line were built, with rail links established as far west as the Mississippi River. Another great building period would occur after the Civil War, and by 1890 the entire continental territory of the United States was connected by 167,000 miles of operating line. Even in the most unpopulated of regions, Americans then lived within close proximity—ten miles on average—of a major transportation system, canal or railroad.

The building of America's great continental transportation network over the course of the nineteenth century served as inspiration for generations of American artists, writers, and folklorists. The transportation revolution has also provided grist for a number of scholarly controversies. One important debate has concerned the very place of transportation improvements in economic development. Some economic historians have specifically depicted the railroads as "prime movers" in industrialization. The railroads, for them, quickly and effectively linked urban and rural communities across the country, dramatically lowering the costs of transporting goods and people and thereby allowing a market economy to emerge. As substantial consumers of raw materials and manufactured goods, the railroads also stimulated the growth of various industries. For scholars writing in this vein, the American past offered a significant lesson to the undeveloped countries of the post–World War II era: modernization and prosperity could be achieved with heavy investments in railroads.

A second and related controversy concerns the question of choices. In the early nineteenth century, Americans debated the wisdom of innovating with different kinds of transportation systems. The merits of canals or railroads occupied discussion. Historians today, using sophisticated statistical techniques, have reconsidered the matter, and their research has placed transportation improvements of a century and a half ago in a new perspective.

The railroads cannot be designated as the leading force behind industrialization. Industrial development preceded the building of the railroads. The emergence of Lowell, Massachusetts, as a major industrial center, for example, cannot be attributed to railroads. Similarly, the railroads did not constitute the primary consumers of any given product; more iron went into the production of nails in the nineteenth century than rails or locomotives. In general, the complexity and unevenness of American industrialization militates against a single cause or "mover."

The railroads also were not necessarily more advantageous than the next best alternative, the canals. The railroads did reduce the costs of transportation; Americans saved money in building extensive railway networks instead of canals. Modern accounting techniques, in fact, prove that supporters of railroads in the nineteenth century had the better of the argument over advocates of canals. Yet, qualifications are in order here too, for the savings have been calculated at about 4 percent of national income; in other words, the graph of American economic growth would have been only slightly less impressive had Americans opted to rely solely on canals for their transport. The crucial matter is that with pressure for expanded market activity and the nation's particular geographic features, major initiatives in transportation were needed; there was nothing determinate in the choice of technologies.

The building of canals and railroads in the United States during and after the Civil War represented an enormous undertaking and entailed massive funding. Here is where government played a definite role. The financing of American industrialization raises a number of curious issues. Americans adopted capital-intensive means of production in a relatively fast and widespread manner. But how was the purchase of plant and machinery possible? There were credit as well as labor scarcities in the new republic. Indeed, the country had a primitive banking system at the time. Attempts to establish a government-sponsored central bank on the order of national banks in Europe foundered on the shoals of political conflict; the short-lived and unpopular First and Second Banks of the United States created by Congress operated under a cloud of suspicion of monopoly power and never fully assumed the role of central bank in regulating the nation's monetary reserves or in bankrolling large-scale enterprises. Private state banks proliferated throughout the country during the antebellum period, and through state-government legislation and private arrangements they brought a modicum of stability to currency matters, but they acted more to facilitate trade through short-term loans than to promote industrial undertakings.

How, then, did American manufacturers purchase the new machinery of the age? The great industrial project at Lowell provides an answer: the gathering together and taking in of partners. Francis Cabot Lowell and Nathan Appleton assembled the financial resources of scores of well-off families from the Boston area who had accumulated fortunes in commerce. Large-scale ventures in the early industrial period involved the pooling of monies, not resort to massive borrowing from banks and deficit financing. When Matthias Baldwin sought to expand his locomotive works in Philadelphia and Cyrus McCormick his reaper factory in Chicago—two other industrial projects that rival the major textile operations of New England—they both followed the path of inviting partners. Their two firms represented pure examples of partnerships, because neither chose to formally incorporate. In some cases, as in New England with the Boston Associates, investors pooled their monies to establish banks to handle their projects, but these enterprises were in some sense partnerships as well.

Plowing back profits represented a second means of affording capital improvements. For small-scale manufacturers in Philadelphia, New York City, and elsewhere, this way toward expansion led through hard work—especially of family members in the case of most proprietorships—and the turning of all profit into new capacities. Growing markets provided the incentive to take risks on investments in the latest technologies.

The financing of canals and railroads presented problems of an entirely different dimension. Thousands of partners would have been required to launch canal or railroad construction projects, and few men of wealth would take the risk of an investment whose return would be forthcoming far in the future, if ever. Private state-based banks in the United States at the time did not entertain loans for such large undertakings. Government was the only agency that could facilitate such funding, and here there were ample precedents. Federal, state, and municipal governments in the country after the American Revolution continued mercantile practices; in the name of the public good, governments offered private groups of businessmen franchise rights and privileges of incorporation to establish public conveniences, including transportation companies.

Government assistance to transportation enterprises was direct and indirect. State officials offered charters of incorporation but also tangible subsidies. An example is provided by the Baltimore & Ohio Railroad, the nation's first major rail carrier. In early February 1827, a group of Baltimore merchants met to discuss their future commercial prospects. The planning and completion of transportation improvement projects in Pennsylvania and New York and the potential diversion of western trade to Philadelphia and New York City had begun seriously to jeopardize their position and prosperity. Among the men assembled, several had followed with interest the results of experiments in England with steam-powered locomotives and rail transport, and they spoke of the possibilities of building a railroad connecting the city of Baltimore with points along the Ohio River. This kind of scene would be repeated in community after community throughout the nineteenth century, with merchants gathered to discuss the need to guarantee continued trade opportunities with the promotion of new transportation linkages.

Within a few weeks time of their initial meeting, the Baltimore businessmen applied to the state legislature of Maryland for a charter of incorporation. On February 28, 1827, an act establishing the Baltimore & Ohio Railroad passed the legislature. The company officially received sole rights to build and operate a railroad along the proposed line, the power of eminent domain, and exemption from state taxation. The state's charter further provided for a capital stock of $3 million, to consist of thirty thousand $100 shares. Ten thousand shares actually were reserved for subscription by the state of Maryland, five thousand for the city of Baltimore. In return for the granting of privileges and subsidization, the state reserved the power to set passenger and freight rates and regulate other aspects of the company's management. This scene in Maryland's state legislature repeated itself in legislative bodies throughout the country as both incorporation rights and financial assistance were bestowed on groups of businessmen who proposed the building of canals and railroads.

All levels of government were involved in the promotion of internal transportation improvements. In the early nineteenth century, the Congress of the United States authorized the building of national roads across states and provided funding. Congress also voted to purchase stock in such state-chartered ventures as the Chesapeake & Delaware Canal Company. In the 1820s, Congress established the practice of awarding federal lands to the states to assist with the financing of transportation projects; the states would raise money through sales of these lands. Congress soon extended the land grant system directly to private companies. From the 1850s through 1880, the U.S. government presented more than 180 million acres of federal land to railroad companies to assist in the construction of railroads through the western states and territories.

State governments played an even more active role in the early industrial period. Between 1790 and 1860, for example, the Commonwealth of Pennsylvania charted 2,333 corporations through special acts of the state assembly. Transportation companies represented two-thirds of the total number of incorporated firms. In legislation establishing canal and railroad corporations, the state subscribed during this time period to no less than $100 million worth of shares. By 1860, Massachusetts had similarly invested $8 million in eight railroad companies, and the state of Missouri had authorized the purchase of $23 million in the stock of its chartered canals and railroads. As the example of the Baltimore & Ohio Railroad indicates, municipalities also played a role. Between 1825 and 1875, 315 cities in New York State had pledged more than $37 million to transportation companies serving their citizenry.

The building of the country's transportation infrastructure during the nineteenth century provides the clearest example of direct government impact on industrial development. But even with this fundamental example, a question has to be raised. How essential? How important was government in the financing of the transportation revolution? Only a mixed answer is possible. In the case of canals, the assistance of government proved critical. The best estimates are that more than 70 percent of the cost of canal construction before the Civil War was assumed by state and municipal governments. The railroads present a different story. Government assistance amounted to 30 percent of all investment in railroads. Private sources of capital were available. Before 1860, foreign investors, in fact, had purchased more than $500 million in railroad securities. The railroads might not have been built as fast without government subsidies, but they could have been established through private means. So popular were the stocks and bonds of American railroad companies that by the 1840s a fairly well-heeled securities underwriting, sales, and exchange system centered in New York City had been established.

The chartering of canal and railroad companies by state legislatures and the financial assistance extended to them by municipal, state, and federal governments are the key pieces of evidence presented by historians who emphasize the essential role played by government in American economic growth and industrialization during the nineteenth century. Direct government initiatives in transportation cannot be ignored, but they cannot be overly emphasized either. The private underwriting of improvements was as important.

For the pre–Civil War period, government in general was not a prime agent of economic development in the United States. The First and Second National Banks of the United States took a backseat to the private state-chartered banks; neither the national or state banks effectively controlled the currency or stimulated enterprise. State chartering of private banks and subsequent regulatory legislation helped create some stability in monetary affairs, yet banknote exchange arrangements privately reached within the banking community proved more efficacious.

As to other direct interventions, the role of government is even less definitive. Along with a transportation revolution, the nation witnessed a communications revolution in the mid-nineteenth century. The trains facilitated the flow of information, but more important was the building of telegraph systems literally paralleling the country's new network of rails, following the invention and perfection of the telegraph by Samuel Morse and others in the 1840s. Despite initial government sponsorship, the spread of telegraphic communication remained a private industry accomplishment. State and local governments similarly made only minimal investments in public education by the Civil War, despite various reform crusades on behalf of common school education; in fact, public financing of schools did not surpass private funding until after the 1860s. Tariff policies remained inconsistent throughout the antebellum period, and the growth of the textile industry in Lowell and elsewhere owed little to the protections afforded by government taxation on imported goods. While tariffs proved important in a few instances—the steel industry can be mentioned—economic historians generally have downplayed their significance in the growth of manufacture at any point in the nineteenth century. If direct government activity had economic impact anywhere, it was probably most critical in the opening of the West to settlement, through western land purchases by the government (the Louisiana Purchase is the principal example here), the seizure of territory through war (especially with Mexico), the herding of the native peoples into reservations, and the subsequent surveying and sale of the public domain to individuals as well as corporate interests.

Why did government not play a greater direct role in economic development in the United States during the early industrial period? Why is government only a small part of the story? The actual record of limited governmental inputs invites these questions, and a key answer is to be found in antimercantilism. The republic had been formed in opposition to the strong, willful state, and suspicion of government persisted. The bases of antistatist belief varied. Southerners opposed government interventions for particular historical as well as ideological reasons. With adequate inland waterway systems, the South did not have a pressing need for transportation improvements and therefore did not welcome the increased taxation required for canal and railroad construction. As importers of manufactured goods, Southerners stood strongly against tariffs. As believers in states' rights, they viewed any expansion in federal power as a threat to their ways and freedom. In the North, on the other hand, antiprivilege and antimonopoly sentiment served as the footing for opposition to increased government. The ghosts of mercantilism and crown corruption hovered. Opponents of government in the North held to the notion that when the state acts, it naturally bestows favor and invariably does so to favored elites.

Interestingly, a laissez-faire critique of government—the position that when the state acts, it inevitably stifles incentive and initiative—did not surface during the early industrial period, despite the writings of influential economic theorists such as Adam Smith and a growing individualist ethos. That argument would not emerge full-blown until late in the century, at a time when a quantum leap in government involvement in economic and social life seemed imminent, and ironically, or perhaps not so ironically, when businessmen were acting in more associative ways. Laissez-faire as an idea can be traced to attacks on mercantilism (the wealth of nations is better advanced through unfettered private enterprise than through government intervention), yet as an ideology, it did not appear until the age of the corporation and would be used by corporate leaders to warn against government interference in their conglomerate operations. The late arrival of the laissez-faire position is one kind of evidence that Americans in the early industrial period remained as republican as liberal, as commonweal-oriented as personally driven.

Support for government intervention to boost the nation's economic might and fortunes certainly existed during the antebellum period—in the pure elitist Hamiltonian and Federalist Party cast and the seemingly more democratic Whiggish and Republican Party mode. Yet, opposition north and south to increased state activity in the economy blunted legislative initiatives, whether on the creation of a strong national banking system, the erection of high tariff walls, or the building of extensive transportation projects.

There was similar contemporary opposition to efforts at legislating morality—to local ordinances banning the sale of alcoholic beverages or requiring the reading of the King James Bible at public events, for example. Cultural and economic issues were in fact intertwined, as manifest in the complex political divisions and alignments of the period. For example, immigrant workers, especially Catholic newcomers, found a home in the Democratic Party, whose populist economic rhetoric appealed to them; in addition, the opposition parties appeared to be dominated by overly pious and zealous Protestants who opposed the immigrants' ways and their very presence in the society. In urban areas, these newcomers quickly learned to use the mechanisms of the Democratic Party to improve their general welfare. Joining immigrant working-class people in the party of Jefferson and Andrew Jackson were a number of other groups: small-scale entrepreneurs angered by the privileges afforded to men of influence, who secured incorporation rights from government; prominent businessmen who supported, for either ideological or personal reasons, free-trade policies; native-born workers of radical persuasion who chose the lesser of two evils when their independent labor party efforts failed; and, significantly, a majority of Southerners, who found the antitariff and general antistatist policies of the Democratic Party conducive economically and who would eventually identify the opposition with crusades to prevent the territorial extension of slavery and even to abolish the institution entirely.

In the other camp, that of the Whig and later the Republican Party, were businessmen who identified with the greater expansion of the American economy through industrialization, western development, government promotion of enterprise, corporate pooling and management of resources, and the empowering of men of merit who could bring prosperity to all. Joining them were older-stock Protestant Americans from all walks of life, especially those from small towns and rural areas, who sought to live in communities marked by reverence and righteous behavior. They believed that moral standards could be upheld through mobilization, organization, and if need be, legislation. For some, this meant participation in religious revivalism; for others, it meant joining movements to strike at social evils in the midst of the republic, be that intemperance or slavery; and for still others, helping to establish institutions—schools, asylums, prisons, and police forces—that could effect greater social order. The Whig and Republican parties were an instrument for those Americans who held that a good and prosperous society could be achieved through the superintendency and stewardship of the upstanding.

Antimercantilist sentiment held sway by and large throughout the antebellum period and blocked attempts at increased state interventions of any kind, particularly in the economy. This stance also left existing

government agencies with few powers and resources, which in turn limited direct state assistance to enterprise. For example, state governments
chartered corporations and maintained regulatory authority over turnpike, canal, railroad, insurance, and banking companies. Yet, inhibited
in both the raising of taxes and the floating of public bonds to amass
monies for state-supported enterprises, these governments could not afford to establish administrative offices for regulation. Fiscal crises, particularly during the depression of the 1830s and 1840s, further left state
governments without any means to subsidize ventures. As a result, state-
created businesses became private enterprises by default. By the 1840s,
the whole notion of founding and empowering corporations to further
the public good dissipated; most states enacted general incorporation
laws at that time, leaving the chartering of firms to administrative agencies and dispensing with the notion of the legislative grant of privilege on
behalf of the commonweal. The commonwealth ideal gave way to private
enterprise, but with widespread suspicion of state power, the state's role
was never sealed or certain during the early industrial period.

Government, then, is only a small chapter in the story of American
industrialization, at least before the Civil War; and the antimercantilist
politics of the period provides the main explanation. Another consideration is timing. While advocates of manufacture tied industrial growth to
the building of a powerful nation-state, the United States began its industrial history not seeking to catch up with already developed and modernized countries (nor was the new republic directly threatened by hostile
neighboring rivals). Late in the nineteenth century and throughout the
twentieth, new nations formed through political consolidations, revolutions, wars, and international treaties have attempted to leap into the
future. The process, supervised from the top, has involved extensive state
sponsorship and coordination of economic development. As one of the
first industrial countries, the United States modernized without pressure
or need for direct government intervention.

The American Political and Legal Systems
and Economic Development

Cataloguing and assessing direct inputs of government to economic
development is only one way of evaluating the state's role in American
industrialization. Another is to look to the general political environment.
Were the political institutions of the society conducive to enterprise? Did
the political and legal structures of the new nation encourage expansive
economic activity? The indirect influence of government may be as im-

portant as, or more important than, specific government interventions. For example, consider the federal system itself.

State governments retained substantial powers under the layered governmental order created by the U.S. Constitution. The authority to charter businesses was one important state prerogative, and a decentralized system of incorporation meant eased access to the privilege. In European countries with centralized political systems, application had to be made to national parliaments, and this required great resources and influence. Because of the ease of access to state approval, the corporate form of enterprise proliferated in the United States to a far greater extent than in Europe. As early as 1800, more than three hundred corporations had been established in the new republic; only twenty such corporate entities could be found in Great Britain at the time. The corporate form allowed for large-scale ventures: vast capital resources could be amassed, and through the gradual legal acceptance of the principle of limited liability, investors confined their risks. This way of conducting business was promoted in the United States by the country's decentralized system of government.

In looking at the ways in which the political structures of the country spurred economic development, attention naturally focuses on the Constitution and its framers. Questions arise. Did the men who participated in the writing of the Constitution in fact deliberately bequeath a system of government that guaranteed economic expansionism? Was the Constitution a "capitalist" document, to put the matter baldly? In considering these questions, care must be taken to distinguish between intentions and consequences.

The intentions of those who were party to the writing of the Constitution remain obscure on all issues. Secrecy marked the proceedings, and no minutes of the deliberations were taken. Surviving recollections indicate vast disagreements, little uniformity, and a great deal of compromise. Nor is there any evidence that the delegates to the constitutional convention imagined the republic to be formed as a mighty economic machine, much less industrialized; they certainly did not connect a decentralized political system with the relatively quick and widespread adoption of the corporate form of business enterprise.

The ambiguity of the intentions of the so-called founding fathers is probably best illustrated in the renowned contract clause of the Constitution. Article 1, Section 10, of the Constitution reads: "No State shall . . . pass any . . . Law impairing the Obligation of Contracts." The inclusion of this clause has been taken as evidence of the framers' desire to construct an unfettered capitalist order. The addition of this article to the Constitution, however, is shrouded in mystery. Extremely late in the proceedings, delegates were considering various prohibitions on state legislative activity (for example, states would not be allowed the right to coin money). Suggestions were made that the states be enjoined from interfering in private contracts, but there was vast disagreement on whether this was or was not to be an absolute proscription. The convention adjourned without a contract clause written. A committee assigned to edit the document as then completed added Article 1, Section 10. Who exactly bears responsibility and the ends in mind were matters disputed at the time and remain in dispute. The intentions of the writers of the Constitution are thus ascertained on this and other issues with great uncertainty.

Intentions aside, the Constitution did establish a particular political order. Was this a framework for expansive economic activity? A more positive document could have been written. The delegates to the constitutional convention, for example, refused, after due consideration, to grant powers to the national government to issue charters of incorporation; to establish national institutes of learning in the arts and sciences; to extend rewards or immunities for advancements in manufacture, agriculture, or commerce; to open canals; to emit bills on the credit of the United States (which would have included notes for circulation as currency); and to make sumptuary laws (taxes on luxuries, for example). The Constitution in general did not lend extensive economic powers to the federal Congress, executive, or judiciary.

On the other hand, various articles in the Constitution did lay a base for a national market system. The most important provision was the authority granted to Congress over interstate commerce; the states would be prevented from placing obstacles in the way of the free flow of goods, people, and currency through the republic. Had the states retained the right to erect separate tariff walls (or to coin money and interfere with contracts), this would have led to a balkanized economy and would not have allowed for regional specialization and, in turn, large-scale enterprise. In creating a federal judiciary that had powers to adjudicate cases between citizens of different states, the Constitution also provided a means to resolve far-flung commercial disputes. A grant to Congress of the right to add new territories and states to the union similarly made possible an expanded market economy.

Other articles in the Constitution would also stimulate market behavior and activity. Congress was given the right to coin money and regulate its value, to establish uniform bankruptcy and naturalization laws, and to create a national postal system with national post roads. The Constitution, in creating a national system of patents and copyrights, further encouraged invention and authorship. The ability of the new central government and state and local governments to borrow money was greatly assisted, of course, by the very establishment in the Constitution of a national state that could assume the funding of all public debt accumulated during the Revolutionary War years. Since states and municipalities would eventually purchase stocks in new publicly chartered canal and railroad companies with monies raised through the floating of public bonds, it was essential to bolster the credit ratings of the young republic, as Alexander Hamilton had argued. (Credit was made available, but also at lowered interest rates as a result.) Finally, in granting federal authorities the powers to assemble militias to suppress insurrections, the Constitution offered forceful protections of private property and interests; the means was there and would be utilized, for example, to quell labor unrest.

Was the Constitution a capitalist document, then? Certainly not by intention. Did the Constitution guarantee economic development? No, again. The emergence of a full-fledged market order and industrialization required a great deal more than the legal sanctification of private property rights, for example. The Constitution did not promote. What it did was *allow*. A political framework was laid down in the United States that placed no obstacles in the way of economic transformation and expansion. The structure of government, then, was conducive to development and was a greater factor, at least before the Civil War, than any particular government inputs.

The interpretation of law is another issue. The Constitution and subsequent legislation can be considered empty vessels. The actual implementation and impact of state policy is significantly shaped by rulings of the courts. In the nineteenth century, the American judiciary—from the highest appellate court of the land to local common pleas courts, where seemingly parochial disputes were arbitrated—occupied a preeminent place in the nation's divided system of government. What role did the courts play in the economic life of the times? More to the point, did the rulings of judges effectively impede or propel industrial development?

On the question of law, the courts, and the economy in the nineteenth century, only a mixed story can be rendered. Bankruptcy laws provide a first example. The Constitution empowered Congress to enact uniform bankruptcy laws. In spite of this directive, Congress moved at a snail's pace on the issue. A national bankruptcy act first passed in the House of Representatives and the Senate in the late 1860s, but this initial legislation was invalidated by various court rulings. A permanent uniform statute would not be put in place until 1893, more than a hundred years after its authorization.

In the absence of congressional activity, local and state courts did seize the initiative on bankruptcy, but a veering course was charted. At first, the courts acted primarily to protect creditors. With heightened political pressure in the 1830s on behalf of debtors and particularly with the abolition of the ancient system of debtors' prisons, legal stances changed. The courts now assumed the role of assisting debtors. In cases of insolvency, the courts moved to keep the indebted in business by establishing schedules of payments to creditors. The courts played a more distinctive role in bankruptcy than the legislative branch, but shifts in judicial approach over the course of the century are notable.

Politics also affected the reading of law on the important matter of contracts. The Constitution proclaimed that no state could pass laws that interfered with obligations assumed and agreed to in contracts by private parties. In 1819 in the landmark case of *Dartmouth College v. Woodward*, the Supreme Court accordingly upheld the inviolability of contracts by declaring that states could not revise charters granted to private institutions. Yet, just eighteen years later—during a period marked by widespread antimonopoly feelings—the Supreme Court in the equally significant *Charles River Bridge* case declared that exclusive privileges could not be granted to corporations and that clauses of original charters could not be deemed as unalterable.

Shifting opinions similarly characterized court rulings in cases involving trade unions, and here, too, politics played a role. From the first indictment against organized labor activity, the Cordwainers case in 1806, local courts into the twentieth century generally held that under common law, labor organizing and strikes represented illegal conspiracies to restrain trade and bring harm. While local courts led the way in the first half of the nineteenth century in limiting collective actions by workers, the federal judiciary later assumed a leading role. Yet, in different places and periods, local judges and juries sympathetic to labor refused to use their offices and verdicts to stifle unions. Organizing campaigns continued despite various legal roadblocks. The right of workers to unionize, strike, and have their representatives bargain with management remained an extremely hazy matter until the 1930s, when definite guarantees were afforded by federal legislation.

If a clear path was blazed by the courts in the nineteenth century with regard to economic affairs, it occurred at the least prominent place in the legal system—at the level of common pleas courts and in civil cases involving private suits. Here, judges tended to favor, even when customary standards were violated, plaintiffs and defendants who were engaged in entrepreneurial activity. For example, the courts absolved factory owners from liability when they diverted waterways to power their mills, actions that often jeopardized the livelihoods of farmers and others who resided downstream. The courts offered other immunities to the enterprising: their customers learned that they had little recourse when they purchased goods that proved to be shoddy; their employees were informed by common pleas judges that in accepting employment, they assumed all risks of the job and that, in cases of accidents, owners could not be held liable for the negligence of supervisors; community members were limited in nuisance claims brought against businesses; and parties to contracts were told that the fine print held more weight than fair dealings. The commons pleas courts in this way contributed to the coming of an unfettered market order.

The notion of public controls on individual behavior was not totally eclipsed, however. Since common pleas courts allowed great leeway in entrepreneurial activity, communities maintained certain regulatory practices: localities continued to control access to trades through licensing laws, uphold quality standards on goods through regulations on weights and measures, and prohibit private actions that threatened the general health and safety of community members. The commonweal principle persisted, and in the late nineteenth century appellate courts sustained municipal, state, and federal regulations of commerce. In the interpretation of law, then, a singular body of rulings did not emerge; the law remained contested.

Limited direct government initiatives, a political structure that left the future open, and legal rulings that bore little consistency: all this adds up to the inconclusive role of the American state in the nation's early economic development. There are readily available answers to explain the less than central part played by government in economic affairs: the antimercantilist politics of the day (with southern and northern variants), the decentralized nature of the American state, the country's early start toward modernization, and the nation's very isolation.

But political conflict itself has to be emphasized, conflict among the first generations of Americans as to the future directions of the republic. Tension existed across two axes—over the extent to which economic activity would be rapacious and expansive and over the place of government in people's lives. The complex of opinion can be schematized as shown in the accompanying diagram. The diagram reveals arguments, not about politics and economics, but about political economy, the combined political economic order of the new republic. Some Americans hoped for a self-regulating society of self-sufficient yeoman producers; others spoke for an America of modern-minded, inventive producers who sought not personal gain but simple well-being and rights for all. Some championed an enterprising, commercialized, individualistic order where merit and not privilege or favor reigned; others yearned for a nation of ordered, ethical communities. A modernized nation-state administered and advanced by civic and corporate elites was hailed by others.

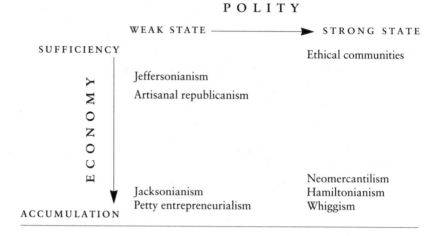

Conflicting Political Economic Views in the Antebellum Period

Antimercantilist politics blocked the formation of a full Hamiltonian or Whiggish state, but lack of consensus itself guaranteed no single way. In some sense, a competitive political and economic order prevailed from the vacuum of discord. Without a common, unified vision, without an effective ideological brake on developments—whether neomercantilist, communitarian, or yeoman producerist—population growth and the nation's very bounty of natural resources ushered in an open, pluralist, unadministered politics and economics by default and happenstance.

The Civil War and Economic Development in the North and South

The Civil War can be envisioned as an end to the impasse. With southern politicians gone from the halls of Congress and the Democratic Party in retreat, Republican Party legislators, identifying with an organized polity and economy, enacted legislation that placed the United States on a definite course. The victory of these latter-day Hamiltonians and Whigs, so the story goes, guaranteed the triumph of an American capitalist system based on the twin foundations of an activist state and corporate enterprise. Does the historical record warrant this interpretation of the role of the Civil War in the nation's economic development?

Debates on the economic impact of the Civil War have generated a great deal of scholarship, and the results point to limited effects. The war certainly did not initiate the nation's march toward economic expansion or industrialization. In fact, in drawing off hundreds of thousands of men into battle and requiring the diversion of resources into the manufacture of military goods, the war actually retarded the rates of production growth established decades earlier. The Civil War also spawned few technological or organizational advances. Parties to the war, especially the North, *utilized* gains achieved in the antebellum period: in machine tooling (and specifically, gun manufacture), in rail transport and telegraphic communication, in factory production, and in financing through the securities markets. The war was more product than agent of industrialization in this respect. If the war had any direct economic impact, it may have been in the area of income distribution. Suppliers of military wares profited immensely, and the war witnessed the channeling of money into the hands of individuals, such as John D. Rockefeller and Andrew Carnegie, who would later invest their gains in the building of mammoth business enterprises of the postwar era. Severe wartime inflation, caused mainly by the massive printing of money to pay for the war effort and by scarcities of consumer goods, actually benefitted debtors and some workers, whose wages rose faster than prices because their skills were in great demand. (Most working people had a harder time making ends meet, and the war witnessed both renewed trade union organizing and civil unrest.) The direct economic impact of the war, with the exception of income effects, must be considered limited; certainly no grand shifts or transformations in economic activity can be delineated.

Historians who have argued for the critical role of the Civil War in transforming the nation economically would note, however, that concentrating on direct effects misses the point. The true impact of the war lies in the political economic realm, specifically in the extraordinary agenda of legislation enacted during the war by northern Republican Party politicians. Among the great initiatives were:

- the Morrill Tariff of 1861, a bill highly protective of American industry;
- the Homestead Act of 1862, providing virtually free land for settlers in designated areas of the West;
- the Morrill Land Grant College Act of 1862, a law granting federal lands to the states for the building of colleges devoted to education in "agriculture and the mechanic arts";
- the National Banking Act of 1863, creating a national banking system and currency;
- several transcontinental railroad bills, offering federal land grants to railroad corporations to subsidize the construction of rail lines connecting the West with the Midwest and East;

- the first federal income tax law;
- a bill creating the National Academy of Sciences to boost technological knowledge and development; and
- further legislation establishing the Department of Agriculture (to
  sponsor research for improved farming techniques), the Bureau of
  Printing and Engraving, the Office of the Comptroller of Currency,
  and the Office of Immigration, the latter all new federal bureaucracies aimed at greater central government leadership in economic
  affairs.

The Civil War thus witnessed the putting in place of a new political
economic order long championed by those who sought to build the
United States into a powerful nation-state through government promotion of growth and large-scale enterprise.

The critical question, however, is whether this "triumph" of a new
capitalist order was more episodic than structural. The latter-day Hamiltonian effort at modernization was not sustained. After the war, the
federal government receded again into the background, and there was a
reversion to localism and unadministered politics and enterprise. The
great nation-building constitutional amendments of the Civil War—the
thirteenth, fourteenth, and fifteenth—which established once and for all
the preeminence of federal edict over state, did not automatically create a
centralized republic; and the South, in spite of its defeat, continued for
generations to block all moves toward a centrally directed economy and
polity. Examples here and there of the impact of Civil War economic
legislation can be cited—research sponsored by the new Department of
Agriculture and information disseminated by the agency to American
farmers bolstered agricultural productivity, and federal land grant policy
assisted large-scale real estate development and railroad corporations—
but the evidence is scattered.

It is thus questionable whether the Civil War actually represented a
substantial shift in the way the country was ordered. Equally questionable is whether a shift toward a market and industrial society in the
particular mid-nineteenth-century American context even required a
necessary capture of government by manufacturing interests and the
writing of a particular legislative agenda. Within three or four decades
after the war, there would be a quantum and permanent leap in the
presence of the central state in the lives of the American people and the
building of a definitively new state corporate capitalist order. This came
in response to great developments and crises of the *post*–Civil War period—the rise of the corporation, pitched conflict between capital and

labor, boom and bust cycles of the economy, and upheavals caused by urbanization and massive immigration—events that had little to do with the Civil War. These developments vitally reshaped the nation, not the Civil War.

This downplaying of the Civil War in our overall political and economic history, of course, completely dismisses (or misses) the South. The war had clear impacts on the South (though even here scholars engage in intense debate over whether the continuities between antebellum and postbellum southern life were greater than the disjunctures). For the entire discussion of government and the economy, in fact, the South tends to disappear from view. Where does the South fit, for example, in the diagram of vying political economic visions and opinion?

One problem in adding the South to the discussion is that although there appears to be a definable region, analyzing and characterizing the South is fraught with pitfalls. There were many Souths: there were obvious intraregional differences (Upper versus Lower South, coastal versus interior); the region comprised a vast population of self-sufficient farm families and small-scale commercial producers, as well as the visible large-scale, slave-worked plantations; and at least before the late 1850s and heightened sectional antagonisms, the region also had its share of Jeffersonians, Jacksonians, and Whigs. As to efforts at defining the South as a political economic entity, scholars have variously described the region as seignorial, prebourgeois, and fully capitalist.

Despite internal differentiation, there are two facts concerning the antebellum South that distinguished it from the North and left it outside certain debates on the future of the republic—debates on whether the country was to be pastoral, a small-producers' commonwealth, highly and petty commercialized, or statist and corporatist. The first was slavery, a production system defended by Southerners as either necessary or ideal. The different visions upheld in the North were all based on notions of free labor.

Second, the South remained a mercantilist offshoot, a region (not quite a "colony") producing staple crops for consumption in outside metropoles and in turn consuming the manufactured goods and financial and shipping services of those centers. Historians have debated whether Southerners (that is, southern plantation owners) were pure capitalists or more aristocrats. (The dynamics of slavery, the dynamics of the slave-master relationship, and the presence of a large poor white class, the argument goes on the latter score, made the planters more lords than bosses.) The fact is that plantation owners managed their enterprises in efficient ways, seeking and earning great profits (although, given the great demand for cotton and the price it fetched on the world market, no matter how they produced they would have amassed the great fortunes they did). Thus, they seem to be capitalists. But the whole debate misses the larger context. The South operated as a mercantilist outpost, not developing an internally based economy with diversified producers and able consumers; indeed, as long as southern exports garnered such wealth, there was little pressure to. This also marked the South from the North, for all sides to political economic debates in the North held aloft the vision of an autonomous republic, one more or less composed of autonomous citizens.

The South, then, is hard to place in discussions of political economic development in the antebellum period. Ostensibly, the South was Jeffersonian. Southerners were for limited government and wary of taxes, tariffs, and (Yankee) corporations; in addition, the modernizers in the North appeared to them to be staunch opponents of slavery. Yet, with the large-scale plantation system and slavery itself, the label "Jeffersonian" fits only in a peculiarly southern way. Slavery and the region's place in the world market economy marked the South, and qualified treatment becomes required.

Separate treatment of the impact of the Civil War is equally necessary; while the northern economy was not affected in substance, the South suffered in distinct and serious ways. With the war fought primarily on southern territory, the region experienced significant losses in personal property and capital resources (barns and railroad stock, for example). With the ending of slavery, Southerners also lost another kind of capital investment. Abolition represented a loss in labor resources as well; freedom for the ex-slaves meant exactly the ability to choose one's hours and pace of toil. (The south would thus experience, with the end of gang-based slave labor, an inward shift of the labor supply curve, to use the terms of economists.) A total reorganization of the production system would be required. Southerners lost in other ways. With cotton production curtailed during the war, consumers of the region's great prize, cotton, looked to develop other sources of supply, namely in Egypt and India. In the aftermath of the war, a glut of cotton appeared on the world market, forcing down prices and the incomes of all Southerners.

Even a cursory overview reveals a far greater economic impact of the Civil War on the South than on the North. Yet, even with the obvious jolts and disjunctures, many historians have been more impressed with how little was changed by the war and during the period of Reconstruction. They point to the extent to which values, social relations, power arrangements, and economic activity within the region remained basically unaltered. There was no "new" South. In comparison with the North during the late nineteenth century, the South does appear relatively frozen in time. Certainly, the South did not experience the second great surge of industrial development—a new spurt also owing little to government intervention—that would mark the nation elsewhere once the last of the battle dead were laid to rest.

Chapter Five

An Industrial Heartland

THE DECADES following the Civil War witnessed an unprecedented expansion in industrial production. In 1860, despite an impressive start in manufacture, the United States still lagged behind Great Britain, France, and Germany in industrial output. Just forty years later, at the turn of the twentieth century, industrial production in the United States would dramatically surpass the *combined* manufacture of its three main rivals. Between 1860 and 1900, American industrialists thus recorded on average a more than fivefold increase in production; the share of manufacture in the nation's total output of commodities grew from 32 percent to 53 percent; and the industrial workforce expanded from 1.5 to 5.9 million workers, who now represented 25 percent of the country's entire labor force.

The industrial ascendancy of the United States in the late nineteenth century raises a number of questions of economic interest. For example, what precisely accounts for the nation's quantum leap in manufacture during the period? If the generative or promoting role of the Civil War is to be discounted, are there other encompassing explanations? Also, did industrialization after the war merely constitute an extension of earlier achievements in manufacture or did the postbellum era mark an entirely new stage of development? Did the United States experience, in effect, a "second" industrial revolution?

Discussion of these questions is forthcoming in this chapter, but first we will focus on another aspect of industrialization in the late nineteenth century—the changing geography of industry. For the antebellum period, a mapmaker would need a fine stylus to fix the location of American manufacture—placing dots here and there, with a few clusters in New

England. For the late nineteenth century, on the other hand, a brush would be in order to stroke a wide swath through New England, the Middle Atlantic states, and the Midwest. The creation of an extensive belt of industry through the greater North and East is an essential component of the story of American manufacture after the Civil War. The new map of industry also highlights a continuing different history for the American South.

Industrial Expansion on the East Coast

A tour of American industry in the late nineteenth century can begin in the East. There, New York City and Philadelphia remained dominant manufacturing centers. New York and Philadelphia more than quadrupled and doubled their populations, respectively, during the period and continued to lead the nation in industrial output. Continuity, in fact, marked manufacture in both cities. Diversified products, diversified work settings, specialization in products and processes, and small to medium-sized, family-owned and -operated firms remained the hallmarks of industry in New York and Philadelphia and continued so far into the twentieth century.

Although a vast array of products issued from New York's and Philadelphia's workshops and factories, one industry—garment-making—occupied a notable place in both cities. The growth in urban populations had greatly increased the demand for inexpensive, ready-made clothes. Increased demand spurred production of standard garments but not, interestingly, large-scale enterprises. Garment manufacturers typically oversaw the cutting of fabrics into patterned pieces and then contracted with lesser operators to complete different aspects of the sewing process. The subcontractors in turn rented space in tenement buildings; utilized a new invention of the age, the foot-pedaled sewing machine, which did not need to be harnessed to a central energy source; and drew their labor from a large pool of newly arrived immigrants from Ireland and England, and later, southern and eastern Europe. Needle trades workers, in fact, would come to form sizable contingents in the workforces of New York and Philadelphia; and the "sweatshops" in which they toiled became sites of labor conflict and grist for social reformers of the day, who petitioned for government regulation of working conditions.

Visitors to Philadelphia and especially New York in the late nineteenth century might have been surprised to learn, however, that these two cities were preeminent in manufacture. Their eyes would have been caught by other activities. The port areas bustled with the traffic of goods and passengers, a reminder that commerce remained a central endeavor. Massive construction of residential, commercial, and industrial properties outward to the city perimeter and upward to a rising city skyscape also drew attention away from manufacture. Finally, the building of new retail emporiums and office buildings in the late nineteenth century made cities such as New York appear to be more centers of information transfer and consumption than of industrial production. The new retail stores and later the offices of the city would also provide new job opportunities for single women as sales and office clerks. These jobs had greater appeal than domestic service and factory work, which remained staples of female employment.

The label "diversified manufacturing centers" could still be affixed to New York and Philadelphia (and Newark, New Jersey, can be included here). In New England, one-industry mill towns similarly maintained their notable place. The fully integrated textile mills of Lowell, Massachusetts, continued to produce staple cloth, but Lowell was now dwarfed by newer textile centers in the region. To the east of Lowell emerged Lawrence, Massachusetts. Developed by Boston investors before the Civil War and concentrating in production of woolens, Lawrence by the 1880s would house three of the four largest textile mills in the country. Bigness marked enterprise in the city. Unlike Lowell, where firms maintained their separate identities, in Lawrence consolidations would occur. The American Woolen Company, founded in 1899, represented an ultimate merger that gave employment to more than twelve thousand workers.

North of Lowell, another initiative of the Boston Associates grew into the *world's* largest textile facility. In 1837, plans were laid for a mill town in Manchester, New Hampshire (the town named for the famed English industrial city). Manchester's planners created only one firm, however: the Amoskeag Manufacturing Company. After the Civil War, the Amoskeag mushroomed to comprise thirty major buildings and seventeen thousand employees. The firm also dominated the social life of the community.

South of Lowell, a third great textile mill center emerged after the Civil War. Fall River, Massachusetts, by the 1870s boasted scores of cotton mills and innovations in technology. The city did not have the benefits of a fast-flowing river for water power; instead, local manufacturers introduced steam engines into their factories and arranged for shipments of coal from eastern Pennsylvania to fuel the boilers. Fall River mill owners also innovated with new, highly automated ring-spinning machinery, though not without incident. Throughout the last decades of the nineteenth century the city experienced serious labor disputes, many of the strikes led by skilled English mule spinners, who felt particularly threatened by the new technologies.

In the shadows of Lowell, Lawrence, Manchester, and Fall River, other smaller New England textile centers could be found. Descendents of Slater-like mills still operated in the countryside of the region; new textile cities such as Woonsocket, Rhode Island, joined the scene in the late nineteenth century. The one-industry mill town thus remained a feature of New England life. There was a notable change, however, in the personnel. A significant aspect of the early history of New England textile production was the employment of Yankee farm girls in the mills. As these young women withdrew from the factories by the late 1840s, the machines of the mills increasingly came to be tended by immigrant workers, both males and females. In the late nineteenth century, successive waves of immigration brought Irish, English, French-Canadian, Polish, Portuguese, Greek, Italian, and Russian newcomers into the communities and mills of New England. Despite deep ethnic divisions and a conservative orientation on the part of some of the new arrivals, mills owners soon learned that they could still not manage their operations without fear of labor unrest.

Changes in the organization of production and not personnel characterized Lynn, Massachusetts, the other great industrial center of New England and capital of shoe manufacture. The late nineteenth century witnessed a definite shift in Lynn toward the centralization of work in factory complexes. The invention of the McKay stitcher played an important role here. Rather than rely on outworkers for the hand-sewing of the upper leathers of the shoes, manufacturers now hired sewers to work in their mills in rooms equipped with banks of sewing machines. The factory workforce grew to include married women (and their daughters, in many instances) as well as single, independent women. Some contracting out of work continued, as did tensions between outworkers and mill hands, married and unmarried women, and male and female workers; but the formation of a larger factory workforce was a notable postbellum development. The participation of men and women in the industry thus continued, but now under single roofs. The workforce in Lynn, especially compared to other New England industrial areas, also remained largely native-born in origin; the city did not receive immigrants in great numbers until the twentieth century. Finally, despite centralization and mechanization, shoe workers still directly handled the tools and materials of their trade—they were not simple machine tenders—and artisanal sensibilities survived. Efforts to forge labor unions in Lynn continued to form a colorful and central part of the city's history after the Civil War.

The mill towns of New England and the diversified manufacturing centers of New York and Philadelphia maintained their notable places on the map of American industrialization, but now visible in the East were other prominent locations of industry. Paralleling the Atlantic coast, for example, was a line of new industrial cities that flourished during the late nineteenth century.

Following the map northward, Wilmington, Delaware, first comes into view. The city began as a grain-processing and distribution center. After the Civil War, the city turned to manufacture and prospered in shipbuilding, railroad car construction, and carriage making. Wilmington made a particular mark, however, in leather tanning, leading the nation in the production of morocco leathers. Access to fresh water, tanning agents, and the international market in hides contributed to the city's prominence in leather manufacture.

Further north, Trenton, New Jersey, was transformed in the late nineteenth century from a sleepy market town and state capital into a major industrial city. Two notable entrepreneurs figured in Trenton's rise in manufacture. Peter Cooper, famous for building the Tom Thumb, one of the nation's first locomotives, founded the Trenton Iron Works in 1857. The ever inventive Cooper developed various new techniques for fabricating iron products, and his firm became a pioneer producer of iron and steel beams and iron facades for building construction.

John Roebling, an engineer by training, not only demonstrated the practicality of building suspension bridges with steel cables, but also succeeded in perfecting the manufacture of metal cable. He opened a wire rope and cable works in Trenton, and by the late nineteenth century the Roebling family firm employed more than 20 percent of Trenton's workforce. Roebling became internationally famous for designing and supervising the construction of the Brooklyn Bridge, a project his son, Washington Roebling, completed after his accidental death at the bridge site.

No famous or single individual figured in the rise of Trenton's other leading industry, pottery products. The area surrounding the city was rich in silicas, clay, and felspar, and enterprising manufacturers utilized these resources in expanded production of ceramic goods, particularly sanitary ware such as sinks, tubs, and toilets. The city also attracted thousands of British immigrants from Staffordshire, England, who lent their skills to making Trenton the pottery center of the country. By 1900, fifteen thousand workers were employed in the forty pottery works of the city.

From Trenton the line of industry next connects to the city of Paterson in northern New Jersey. Paterson had been slated for industrial prominence since the days of Alexander Hamilton and the Society for the Establishment of Useful Manufacture. Little came of this original venture, but by the time of the Civil War, the city housed successful locomotive works and cotton mills. Paterson's great ascendance, however, occurred after the war, as it emerged as "Silk City," the leading center of silk textile production in the United States. Paterson benefitted from supplies of fresh water for the dyeing of the silk, an existing machinery industry, and close proximity to the fashion markets of New York City. Events in England, though, propelled Paterson's silk future. Reductions in tariffs on imported silk products decimated the British silk industry in the 1860s. Thousands of skilled British silk workers then fled across the Atlantic to develop and staff Paterson's growing silk trade. By 1900, the silk mills in the city provided employment for more than 20,000 residents. Nearby, another textile center emerged in Passaic, New Jersey, although here woolen mills predominated.

The line of new industrial centers paralleling the east coast bends further into Connecticut. The city of Bridgeport on Long Island Sound sat at the base of a valley of industry. The city itself became famous by the early twentieth century for the manufacture of specialized metal products, particularly machine tools, rifles, and ammunition casings. North of Bridgeport a band of towns had emerged, centered around Waterbury, Connecticut, that served as the center of brass and brass product manufacture in the United States. Copper and zinc were melted in the mills of the area, and the resultant alloy was then formed into ingots, sheets, wire, and tubes for the fabrication of various wares from brass buttons to kerosene lamps. Clock and watch parts were other important products of the brass works, and the so-called brass valley of Connecticut also became the capital of American clock and watch manufacture.

Farthest east of the new industrial cities of the Atlantic seaboard was Providence, Rhode Island. Providence had long played an important role in the nation's commerce, but after the Civil War the city emerged as a proud center of manufacture. A vast array of goods poured from the factories and mills of the city, but for a few products, Providence made distinguished contributions. Tools from the Brown and Sharp Company, files from Nicolson's, steam engines from Corliss, and flat silverware from Gorham's were respected around the world. Providence also led in woolens production and was the nation's capital for jewelry manufacture.

Worcester, Massachusetts, inland and not located on a riverway, was an unlikely place for a major industrial center. However, the completion of railroad connections and the steam engine allowed for the success of its manufactories after the Civil War, and it joined the list of prominent northeastern industrial places. Product diversity marked Worcester's manufacturing development, but more than 40 percent of the city's workforce could be found in the metal trades producing machinery, tools, and metal wire. In a state dominated by textile towns, Worcester stood alone.

The industrial cities that emerged in the late nineteenth century along

the eastern coast shared a number of features. The cities, stretching from
Wilmington to Providence and Worcester, were diversified manufacturing centers. In this respect they had more in common with New York
City and Philadelphia than with the textile towns of New England. As in
New York and Philadelphia, the firms in these cities engaged in the small-
batch production of specialty goods, and family-owned and -operated
enterprises prevailed. (With few exceptions, companies in these cities
were not involved in the merger movement of the late nineteenth century
that led to the formation of nationally based corporations.) But unlike
New York and Philadelphia, these smaller cities had dominant or hall-
mark industries—whether it was leather in Wilmington or jewelry in
Providence.

Skilled immigrant workers also played important roles in the success
of manufacture in these cities: there were the potters from Staffordshire,
England, in Trenton and the brass workers from Birmingham, England,
in Waterbury. The process whereby skilled European workers contributed to the transfer of technical knowledge and expertise from Europe to
the United States continued after the Civil War.

Finally, these new industrial centers, unlike metropolitan New York
and Philadelphia, were relatively enclosed worlds. Local manufacturers
in the cities formed discernible elites and played leading roles in civic
affairs. Through the force of their authority, but also through their extended goodwill and respect for republican values (many had risen from
the shop floor), they affected social order. For this and other reasons, the
new manufacturing centers of the east coast did not experience as fully
the great battles between capital and labor that were to unfold during
this period.

The Western Expansion of Industry

The filling in of the map of industry along the east coast is easily
overshadowed by the dramatic opening after the Civil War of an industrial heartland to the west, stretching to Chicago. A first line of western
development can be traced across New York State. Here, along a 350-mile
corridor that embraced the Erie Canal, a chain of industrial cities emerged.
Most had been market and shipping centers before the war and then
flourished afterward in manufacture.

Access to iron deposits and wood supplies allowed Albany, in the east,
to emerge as an iron-manufacturing center. Cast-iron stoves became the
specialty of Albany's iron foundries, but rolling mills in the city produced
impressive amounts of nails, rails, and iron plate for vessels. Albany

prospered as well in brewing, meatpacking, tanning, textile machinery, and flour milling and served as the location of the massive repair shops of the New York Central Railroad.

Albany formed an industrial complex with two other nearby cities— Cohoes, a cotton textile center, and Troy, renowned in iron production. With the advantage of extensive water power, iron manufacturers in Troy assembled more expansive facilities than existed in Albany. Just west of Albany, Schenectady developed as a manufacturing center with the founding of a major locomotive works. Thomas Edison later opened an electrical machinery plant in the city which would evolve into the massive manufacturing headquarters of the General Electric Company.

Ninety miles west of the Albany complex, Utica etched out a place in industry, first in textiles, but then more notably in garment manufacture. What distinguished Utica was the centralized production of apparels in large steam-powered factories. Nearby Rome, New York, at the same time emerged as "Copper City," foremost center for the production of copper products, including, most importantly, copper wire.

Industrial development along the Erie Canal skipped over Syracuse, in the middle of the corridor; Syracuse remained primarily a market center, although the processing of salt from surrounding salt flats became a considerable enterprise. A number of smaller manufacturing towns emerged west of Syracuse, but glaring on the map were the two major industrial cities of Rochester and Buffalo on the western ends of the chain.

Rochester had been a boomtown before the Civil War; while interest in American manufacture during the antebellum period focused on cities to the east, Rochester's flour and lumber mills, machine shops, barrel works, and garment manufactories turned out products in sufficient number to place the city at the upper ranks of industry by 1860. The postbellum period was marked by consolidation and expansion of existing facilities. Large, centralized garment and shoemaking plants dominated the city's landscape. Rochester's fame in manufacture, however, would rest and flourish on the inventions of one of its leading citizens. By the mid-1880s, George Eastman had perfected a small hand-held camera for popular use (the Kodak) and film for it, and he would oversee the establishment in Rochester of the world's largest manufactory of photographic equipment.

Buffalo, west of Rochester, boasted little industry in 1860. The city was a major transshipment point between the Great Lakes, the Erie Canal, and places east; and commerce, shipping, and grain storage were chief pursuits before 1860. In the thirty years after the Civil War, Buffalo emerged as an industrial giant. Although various industries found a place in the city, iron and steel production defined Buffalo's future. With easy

access to coal deposits in Pennsylvania and the iron ore of the western
Great Lake states, Buffalo was perfectly situated. The iron and steel mills
of Buffalo also required huge labor resources, and the city, like other manufacturing centers of the Erie Canal corridor, attracted hundreds of thousands of immigrants during the late nineteenth century. Irish and German
immigrant workers, in particular, helped build this line of industry.

Another tier of manufacture could be delineated across the southern
boundary of New York State, with centers of shoemaking, glass blowing,
and railroad construction and repair. But more prodigious developments
were unfolding further south in the state of Pennsylvania. Pennsylvania,
throughout the nineteenth century, remained the leading industrial state
in the nation both in terms of employment and the value of goods produced. Rather than lines, large circles are needed to mark the place and
extent of industry in Pennsylvania. At the southeastern corner, metropolitan Philadelphia would be highlighted, for that city continued to rival
New York as the country's leading manufacturing center. North of the city
another great circle would be employed to designate an area critical to
the country's entire industrial progress.

Below the surface of a five-county region beginning some fifty miles
north of Philadelphia lay 95 percent of the nation's supply of anthracite
coal. As early as the third decade of the nineteenth century, investors
from Philadelphia were already betting on the value of this resource.
They sponsored first canal and then railroad construction into the area
to facilitate the transport of the coal to urban centers. Through their
transportation companies, they also began purchasing enormous tracts
of land to monopolize claims to the fossil fuel. Holdings were then leased
at great profit to small-scale operators or mined directly by the companies. With the opening up of the anthracite coal fields, other independent producers also tried their hands. By the time of the Civil War, mine
shafts had thus been quickly (and hazardously) excavated throughout the
region, but the massive exploitation of the area's great resource awaited
war's end.

The importance of coal in American industrialization cannot be over-
stated. A more efficient fuel than wood, coal allowed for expanded use of
steam engines in transportation and manufacture. Industrial plants were
no longer bound to water sites, with steam engines and coal in this way
promoting the geographical expansion of industry. Steam engines also
permitted large-scale manufacturing, which in turn fostered corporate
development. There is thus a direct connection between coal, mass industry, and the rise of the bureaucratic corporation. Coal could also be
transformed into other products, and various derivatives of coal proved
essential in industrial progress. For example, coke, a distillate of coal,

was the required fuel in steelmaking. Burned coal produced methane gas, which was employed in street lighting. Finally, anthracite coal, in particular, burned slowly and cleanly and had significant use in home heating.

Expanded demand for coal in the late nineteenth century brought boom times to the anthracite region of eastern Pennsylvania. Small cities like Pottsville emerged, handling the outward and inward flow of coal and consumer goods. Larger cities, such as Scranton and Wilkes-Barre in the far north, appeared as well, first as coal mining and coal transportation centers, but then attracting and fostering other industries, including textiles. In between these cities could be found scores of mine patch communities and company towns. The bringing of coal up from deep in the earth required enormous labor power, and to the region came a succession of immigrants from England, Ireland, and Wales and then southern and eastern Europe. The exploitative conditions under which they worked and traditions of independence and protest among miners generated trade union campaigns and ongoing confrontations between workers and owners that proved to be among the most dramatic of the age.

Intense labor conflict would also mark Pennsylvania's third great industrial district. On a map of industry, most of Pennsylvania would be left blank, the central and western portions of the state remaining largely farm and forest land. A thin line could be drawn across its southern boundary designating the machine-shop and textile town of Reading, the iron and steel works of the Harrisburg area, and notably, the mammoth repair complex of the Pennsylvania Railroad in Altoona. It would be at the western end of that line that a major tag would need to be placed, for there, in Pittsburgh and its environs, emerged an emblematic American late-nineteenth-century hub of industry.

With nearby rich deposits of coal, iron ore, and other minerals and its water and railroad linkages in all directions, Pittsburgh was primed for success. By the onset of the Civil War, the city had already become the leading glass-producing center in the country. Glass and rail transport remained key endeavors after the war, but the city's new fame would be based on iron and steel manufacture. By the turn of the twentieth century, with the city accounting for one-sixth of the nation's output of iron and steel, Pittsburgh and steel became synonymous.

The picture of Pittsburgh in the mind's eye is of sprawling steelworks, powerful corporations, and masses of workers. The reality is more complicated. By the 1880s, numerous ironworks had been established in the city and in surrounding communities in Allegheny County. These tended to be proprietorships of 200 to 300 employees. Iron mill owners generally left the management of their works to skilled ironworkers, who supervised teams of men whom they often hired directly. Skilled puddlers oversaw the difficult mixing and heating of the ore; rollers, the shaping of the molten iron into ingots, sheets, and rails; molders, the preparation of casts; and forgers, the hammering into shape of large iron components. Mill owners reached per-ton and per-piece agreements with the skilled men, and as these craftsmen organized into unions, arrangements began to be negotiated on a collective basis.

The iron craftsmen's hold on production began to be challenged in the late 1880s, but with an attack on many fronts. There was a general move toward greater production of versatile steel and in Bessemer converters, which eliminated the skills of the puddlers. Business consolidations occurred, reducing competition and also allowing for investments in new capital-intensive technologies. Mechanical moving devices and other automated processes then further reduced skill needs. New corporate enterprises also began to hire newly arrived immigrants from southern and central Europe at low wages to tend the new machinery, and they were supervised now by salaried bureaucrats. Finally, and most important, steel executives deliberately moved to break the power of the strong craft unions in the industry. The defeat of the unions in a series of dramatic strikes in the late 1880s and early 1890s boosted managerial control over operations. The violent, monumental Homestead strike of 1892 capped developments. The city of Pittsburgh became headquarters to such corporate giants as the United States Steel Corporation (formed in 1901), but the great steel combines of the city—corporately owned, bureaucratically managed, mechanized, and automated—emerged only after a long unfolding history and amid great conflict.

If large circles are required to represent the location of industry in Pennsylvania—in Philadelphia, the anthracite coal fields, and Pittsburgh—a great splash of points would be appropriate for Ohio, a prime agricultural state that was actually awash with industry by the late nineteenth century. In 1880, 60 percent of the state's working population could be found employed in its widespread manufactories. There were a number of reasons for Ohio's flourishing industry. Coal and iron resources proved a boon; natural and manmade transportation networks fostered development; and the state's mushrooming population greatly boosted demand for manufactured products. The range of industrial centers in the state is also impressive. Ohio had major industrial cities, such as Cincinnati and Cleveland, but also a large middle tier—from Akron to Toledo—and numerous small manufacturing towns and mining communities.

Cincinnati had emerged as a capital of industry before the Civil War. By 1860, remarkably, the city had achieved third position behind New York and Philadelphia in manufacturing output. Cincinnati shared features with its eastern rivals: a diversity of products and work sites, specialization, and small to medium-sized family-owned and -operated enterprises characterized its industrial structure. An impressive range of goods flowed from the city's manufactories: furniture, wagons, carriages, coffins, plug tobacco, beer, whiskey, hardware, machinery, boots, shoes, clothing, and soap. The list is slightly different from comparable compilations for New York and Philadelphia. Resources shaped Cincinnati's specific manifest of goods—for example, local supplies of wood allowed for expanded manufacture of wood products—as did demand from Ohio's growing population of farm families, whose engagement in commercialized agriculture made them reliant on the marketplace for needed commodities. Cincinnati's specialized workshops also required skilled labor; and like Philadelphia, the city attracted sizable numbers of skilled immigrant workers, particularly from Germany.

Cincinnati could boast of an array of products, but one industry garnered special attention, and that was meatpacking, or to be more precise, pork packaging. At an early date, the city earned the title "Porkopolis" for the number of pork slaughterhouses operating within its boundaries. Of great intrigue was the assembly-line technique of butchering and producing pork products that was developed by Cincinnati's packers. The fat renderings also allowed for the growth of an important soap industry in the city; Cincinnati would thus be headquarters for the famed Proctor & Gamble Company, founded in the 1830s.

As with New York and Philadelphia, continuity largely marked Cincinnati's manufacturing history after the Civil War. The city remained a diversified industrial center. Factory size increased, greater mechanization and division of labor diluted skill levels, and large corporately owned enterprises made their appearance—all to greater extent, in fact, than in New York and Philadelphia—but still small-scale producer ideals persisted and affected civic affairs.

Cleveland, Ohio's other major industrial center, mostly resembled Buffalo, its Great Lakes rival, and to some extent Pittsburgh. With transportation links in all directions through canals, lakes, and railroads, the city had emerged as an important transshipment point at the time of the Civil War. Industrial expansion in clothing, brewing, meatpacking, machine tools, and shipbuilding would mark its postbellum development, but Cleveland then would become chiefly identified with iron and steel production and oil refining. Accessibility to iron ore and coal reserves allowed Cleveland to join Buffalo as a major center for iron and steel manufacture and the eventual site of large-scale steelworks. It would be oil discoveries first in western Pennsylvania in the 1850s and later in Ohio that would present special opportunities for Cleveland. The city was ideally situated for receiving crude oil by railcar and pipeline and then for the processing of oil products and their distribution. In addition to the city's definite locational advantages, serendipity also played something of a role. John D. Rockefeller had moved with his family to Cleveland in 1854; he began in business there as a produce merchant, and then, with discoveries of oil, he made investments in refineries. By the 1870s he would come to dominate the oil shipping and processing industry and would oversee the building in Cleveland of the nation's largest oil refinery.

Cincinnati and Cleveland would be joined by a host of smaller cities in contributing to Ohio's total industrial product. All of these cities were characterized by diversified manufacture, but some came to be identified with particular goods. Canton would be known for watches; Springfield for agricultural machinery; Youngstown for steel; Toledo for wagons, steel, and glass; Dayton for railroad cars and office machinery; and Akron for rubber. Even smaller places gained acclaim. East Liverpool, Ohio, for example, emerged as a pottery center to rival Trenton, New Jersey; the town had an appropriate name, for it was populated largely by skilled English immigrant potters. Finally, dotting the southeastern portion of the state were scores of bituminous coal mining villages.

The path of American industry then passed widely through Ohio, but it would skip over Indiana. There was budding industrial activity in Fort Wayne and Indianapolis, and early in the twentieth century, major manufacturing centers would emerge in the state—the immense steelworks at Gary being of greatest note—but while industry surged through neighboring Ohio, Indiana remained primarily agricultural.

Industry touched Michigan with slightly greater force. Detroit would not join Buffalo, Pittsburgh, and Cleveland as one of the new industrial giants of the age until the first decades of the twentieth century, when the city became the capital of automobile manufacture for the nation and the world. Still, by the 1890s Detroit assumed its place on the map of industry as a diversified manufacturing center. The city's growing, largely immigrant population found employment in railroad car construction, iron mills, machine shops, pharmaceuticals, shoemaking, and seed processing. Detroit actually shared honors at the time in manufacture in Michigan with the city of Grand Rapids, located to the west. Drawing from the great timber resources throughout the state, Grand Rapids emerged as early as the 1860s as the leading furniture-producing center in the country.

West and north of Ohio, industrial designations are thus visible for Detroit and Grand Rapids and such cities as Milwaukee, Wisconsin— that city earning great fame after the Civil War for the manufacture of beer, among other products. But with the map of midwestern industry in view, the eye is automatically drawn to a greater marker placed on the southwest shores of Lake Michigan. There rose Chicago, an urban center to rival New York, a city epitomizing American industrialization in the late nineteenth century.

Trade constituted the initial base on which Chicago's fortunes rested. The immediate site on the southwest shores of Lake Michigan where the city was founded was not particularly advantageous. The area was swampy and not served by notable inland waterways. Drainage posed such a problem that in 1849 the buildings and streets of the emerging metropolis had to be jacked up and placed on pilings. Developers, though, saw beyond the proximate problems. As early as the 1830s, real estate speculators realized the extraordinary potential of the location. To the northwest, west, southwest, and southeast lay America's agricultural future. Chicago would be the receiver and distributor of the bounty of its wide hinterland. And due east lay population centers to which these goods could be shipped. In anticipation, then, of hinterland development, Chicago's founders outlined a city, began a frenzied selling and purchasing of lots, and promoted natural waterway improvements and canal and railroad construction. Emanating from the city there would be a fan of waterways and railroads linking Chicago with its agricultural surroundings and with Great Lake shipping lanes and major railroad trunk lines connecting it with eastern markets.

The receiving, processing, and marketing of the products of Chicago's hinterland shaped the very physical layout of the city, too. Huge grain elevators appeared where railroads entered the city in the south and west. Acres and acres of rail freight yards emerged alongside. On the shores of Lake Michigan, cargo areas were built for the boatloads of lumber arriving daily from the north.

In the southwest, Chicago's prominent meatpackers allied to construct the city's famous (or infamous) stockyards. Cattle and pigs, transported to the city by train from prairie states, were held there ready to be killed and dismembered in adjoining slaughterhouses. In the 1860s, Chicago's meatpackers had invested millions of dollars in the building of large, mechanized packinghouses to facilitate a mass processing of meat never contemplated before. The investment was based on their assessment that they would have access to huge supplies of cattle and pigs, bought cheaply in volume; that they could employ efficiently tens of thousands of workers in the killing and disassembly of the animals; and that they could easily market their cut meats in eastern markets. Innovations in refrigeration, advertising, and merchandising aided in the last effort. Chicago thus became the greatest meatpacking center in the world.

Trade also promoted the building of an impressive downtown of office buildings and retail outlets. The massive transfer of goods in the city required banks, insurance companies, and mercantile exchanges where contracts for current and even future deliveries of agricultural products could be bought and sold. Merchants of all kinds also plied their wares in the downtown, and visitors from the countryside on annual visits could buy gifts and fancy goods at several of Chicago's famed emporiums. Chicago's merchandisers even perfected the means to sell goods to rural families who could not venture to the metropolis; in the late nineteenth century, Chicago would house the two great mail-order enterprises of their day, Montgomery Ward and Sears, Roebuck.

Economic activity in Chicago centered on the processing and marketing of the resources of its vast and bountiful hinterland and in servicing its rural neighbors near and far. Employment opportunities were to be found in Chicago along the docks and in the railroad yards, but also in industries involved in the processing of the products of the surrounding land and forests: in meatpacking, lumber milling, flour milling, tanning, and soapmaking. After the Civil War, however, the city began to develop a strong new base of manufacture. Iron reserves in nearby western Great Lake states and coal from downstate Illinois stimulated iron and steel production in the city; and by the turn of the twentieth century, Chicago's massive steelworks rivalled those of Pittsburgh, Buffalo, and Cleveland. Manufacturers did not try to compete with eastern producers of textiles, but they could take advantage of a huge urban and rural market for ready-made clothes. A number of the largest centralized clothing factories in the country would emerge in the city.

Industrialization in Chicago in the late nineteenth century is perhaps best identified with the formidable McCormick reaper works. Cyrus McCormick succeeded in the development and manufacture of reaping machines before the Civil War. In the 1850s, he wisely decided to make Chicago the headquarters of his operations, and over the next thirty years the company greatly expanded its facilities in the city. By the mid-1880s, employment reached 1,500 men, and production, more than 50,000 reaping machines a year, far greater than the company's nearest competitors. As was the case with iron mills in Pittsburgh, executives at McCormick attempted to achieve increased efficiencies through extended mechanization, division of labor, defeat of craft unions, and increased supervision. This fomented conflict with skilled workers, who had maintained great control over production. One critical confrontation between McCormick workers and management would lead to the great Haymarket Square riot of May 1886.

With the McCormick reaper works and other large-scale industrial enterprises, Chicago emerged in the late nineteenth century as an industrial as well as a commercial colossus. Young men and women flocked to the city from the surrounding countryside seeking their fortunes—a migration which produced both rural anxiety and a new genre of literature—while hundreds of thousands of immigrants, notably from Germany and central Europe, also arrived seeking opportunity. By the first decade of the twentieth century, Chicago's population swelled beyond 1.5 million, and the city was now second only to New York as the nation's greatest metropolis.

Chicago grandly anchored the western end of a wide belt of industry that covered New England, the Middle Atlantic states, and the Midwest. There were separate configurations—lines of industry along the Atlantic coast and across upstate New York and pockets of enterprise in Pennsylvania—but the whole represented America's claim to industrial supremacy by the turn of the twentieth century. There would be manufacturing development to the west of Chicago, but in some sense the centers that did emerge were annexes of Chicago. Minneapolis became a capital of grain storage and processing, Omaha and Kansas City of meatpacking. With transportation improvements, these cities were closer to opening agricultural areas than Chicago and assumed many of Chicago's functions. Other industrial sites similarly appeared for the manufacture of farm machinery, such as Moline, Illinois, and Davenport, Iowa. The expansion of the railroad system also fostered the development of new marketing centers in the West, and mineral discoveries in the mountain states gave rise to other western cities. These developments, however, occurred as extensions of the industrial core of the country that stretched from Lynn, Massachusetts, to Philadelphia and west to Chicago. The United States would see the creation of one more belt of industry, along the Pacific coast states, but that would not occur until the twentieth century and would pale in comparison to the industrial heartland fashioned a century earlier.

Postbellum Economic Development in the South

The American South seemingly does not enter into this story. There was industrial growth in the South after the Civil War, but developments in manufacture there are overshadowed by a general economic and social history that varies distinctly from the rest of the country.

In 1860, the American South was one of the wealthiest regions in the world; economists estimate that had the South seceded without resistance, it would have joined the ranks of independent nations as the world's fourth richest country. The great income that the South derived from the sale of cotton at home and abroad, of course, was not widely shared among its inhabitants, but income distributions in region were not much more inequitable than in the free-labor North. The decades following the Civil War found the South in different circumstances; with war's end, the region entered a long-term period of economic stagnation that stretched through World War II. The last decades of the nineteenth century witnessed growth in the South, but relative to the rest of the nation, the region lagged far behind and appeared to be a backwater.

There is no dearth of explanations for the South's new straits. Southerners pointed to the destruction wreaked by the Northern armies. A glut of cotton on the world market and falling prices certainly took their toll. The South suffered too from its antebellum success; with energies and monies fixed on the growing of cotton, the region emerged after the war without an adequate infrastructure of commercial, financial, transportation, and manufacturing facilities, and Southerners remained reliant on outside purveyors of goods and services as a result. A weak banking system, in particular, left the South critically short on credit and currency. (The latter, it was argued, made the move from slavery to a wage-based economy difficult.) The South's banks were further penalized by strictures of the new national banking system created by the National Banking Act of 1863. The list of causes can be expanded, and all of the above answers have merit. Still, the major cause for the South's post-bellum economic decline lay in the rearrangements of production that were forced by northern victory and the abolition of slave labor.

Southern landowners at war's end had hoped to plant and harvest their fields with the use of teams of wage laborers (in lieu of slavery). In mobilizing their workforces, the former plantation masters received assistance from the Freedmen's Bureau, an agency created by Congress to aid former slaves in their transition to freedom. Eager to prove the benefits of the wage system and disturbed by the seemingly lax behavior of the freedmen, bureau officials pressured former slaves to sign contracts pledging their labors to specific employers. Bureau agents, however, had no better success than the landowners in persuading the freedmen to return to the fields. In general, African-Americans in the South after the Civil War refused to work the land in servile ways. Wage labor offered few securities to them, and gang work brought back memories of slavery. Most important, though, the freedmen wanted to till their own homesteads; in fact, promises had been held out that the federal government would confiscate plantation land and award families of freedmen "forty acres and a mule."

A stalemate prevailed—with landowners needing labor and land workers wanting land—and out of the standoff emerged a negotiated settlement: sharecropping. Landowners divided their lands and reached agreements with individual freedmen to work the separate parcels. Freedmen thereby gained access to the land and control over the pace of work and their daily lives. In return, they owed the landowners a portion of the harvested crops; while there was some room for bargaining, a fifty-fifty divide was the normal arrangement. For their part, landowners had hands to work the fields, no greater obligations to them, and crops to sell. By itself, the sharecropping system did not foster stagnation and poverty. Landowners and sharecroppers shared in the risks of production, and technically, both sides could have reacted wisely to the cues of the marketplace and agreed to a crop mix that guaranteed the greatest returns to all.

Layered over the sharecropping system were other arrangements and factors that made for misfortune. Sharecroppers required credit. They incurred costs over the course of the whole year—needing to purchase seed, tools, food, clothing, and other wares—but received compensation for their labors at only one point, the harvest time. After the Civil War new merchants—both independent merchants and landowners who assumed the function—appeared in the southern countryside to offer credit to the region's sharecroppers. Sharecroppers invariably acquired significant debts, especially in circumstances where single merchants monopolized control over credit and fixed interest rates almost at will. Sharecroppers at harvest time then not only owed substantial portions of the crop to the landowners, but also owed the merchants. The croppers frequently discovered that they were left with little return of their own after paying their creditors and even that they were in arrears. Merchants would carry over the balance of debts into the new year. Sharecroppers throughout the South thereby entered into cycles of perpetual debt.

Sharecropping and debt peonage had grave consequences for the individuals involved as well as for the greater southern economy. The merchants' very survival rested on keeping their clients dependent on them. They insisted that the sharecroppers grow only cotton and no crops or livestock that would allow for self-sufficiency (and lessen reliance on the local storekeepers). The merchants, too, stayed with cotton, a tried and seemingly true income producer. Because there was a glut of cotton on the world market, the dependency relationship between merchants and sharecroppers bound both groups to a commodity whose earning power was declining, and they were unable to react to the messages of the marketplace and switch to other crops. The South became locked in to a faulty economic program.

The sharecroppers also became locked in place in personal ways that had broad ramifications. Because they were in debt, they could not move—although family survival often forced the male heads of households to travel at times during the year to find extra work in mining, timbering,

railroad construction, and hauling. Here race relations and racial hostili-
ties in the region played a role as well; through threatened or actual
physical intimidation and economic pressure, African-Americans after
both the Civil War and the failed experiments in political and civil rights
of Reconstruction learned to keep their place. Sharecropping involved
both poor white and poor black farm families, but racism functioned to
further immobilize the black population. The freedmen could not wan-
der at will seeking opportunity and the best situations, and consequently,
a truly free labor market was not established in the South with the
abolition of slavery. This, too, distinguished the South from the North.

This immobilized, isolated labor force was also a key ingredient in the
South's dire economic circumstances after the Civil War. Labor should
have flowed out of the region to higher wage frontiers. With labor kept in
place by land and credit arrangements and racial practices, the South had
a surfeit of low-paid workers, which was advantageous to employers in
the short run, but not advantageous for the South's economic future.
Unlike the North, the South did not face the kinds of labor scarcities that
would force capital development. One lumber mill operator in the late
nineteenth century neatly encapsulated the point with this statement:
"Instead of installing machinery to do the work, we always undertook to
do it putting in another cheap negro."

In the late nineteenth century, there were Southerners who understood
the implications. The South could not prosper with a low-wage agricultural system. The only solution lay in industry, and the region had its
legion of vocal advocates of manufacture. The most notable was Henry
W. Grady, editor of the influential newspaper the *Atlanta Constitution*.
In editorial after editorial, Grady championed industry, even arguing that
Southerners should openly welcome northern investors. In an oft repeated quote, Grady chastised Southerners for their unprogressive ways.
Describing a funeral and the goods he espied at the event, Grady wrote,
"The South didn't furnish a thing on earth for that funeral but the corpse
and the hole in the ground." Grady echoed sentiments voiced earlier by
J. D. B. De Bow, editor of the well-read *De Bow's Review*, who declared
that industry was the "true remedy" for southern ills. "We have got to go
to manufacturing to save ourselves."

The campaign for industry had a direct impact. In the last decades of
the nineteenth century, the South began to build an important belt of
industry of its own. Across the Piedmont—an area of rolling hills and
rushing waters stretching from southern Virginia through central North
and South Carolina and into northern Georgia and Alabama—emerged
a region of textile mill villages. Cotton textile development had begun in
the Piedmont before the Civil War, but in the 1870s and through the

1880s, a surge in mill building occurred. Local investors—planters, merchants, and professionals—sponsored initiatives in the name of providing employment for struggling farm families and lifting their communities out of poverty. "Next to God, what this town needs most is a cotton mill," one local booster proclaimed; and promoters in such spots as Gastonia, North Carolina, and Spartanburg, South Carolina, placed their own reputations and that of their towns on the line in seeing to the building of local mills. This was an indigenous movement; it would only be during and after the 1890s that northern textile companies began looking to move their operations to the South to take advantage of the region's low-wage, nonunion labor base, a migration that would contribute to the collapse of the northern textile industry (though not fully until the mid-1920s).

Of equal note were the workers who came to toil in the textile mills of the Piedmont. Mill owners deliberately recruited poor white farm families. The post–Civil War period saw the penetration of market activity into backcountry regions throughout the South; railroad expansion and the spread of merchant stores hastened developments. Whether working now as sharecroppers or tenant farmers or still eking out livings as self-sufficient producers, the South's white yeomanry still faced hard times. They thus represented a source of labor for industry, and mill owners attracted them to the mill villages with promises of work: for the women and children, positions as machine tenders, and for the men, construction, maintenance, and other odd jobs. Poor black families were not afforded these opportunities, and the mills remained lily-white decades into the twentieth century.

Just west of the South's textile district, another belt of enterprise emerged, though this was not so directly the result of efforts by local boosters and others who saw industry as the great panacea for the region. The mountain areas of West Virginia, eastern Kentucky, eastern Tennessee, and northern Alabama contained rich deposits of coal and iron. In the late nineteenth century, this region would be opened to mining, which gave employment to Appalachian mountain whites as well as to black men who migrated from surrounding areas to find part-time and full-time work.

The iron ore and coal of the southern Appalachians in turn served as the basis for a new southern iron and steel industry. Before the Civil War, the South had only one major iron mill, the Tredegar Iron Works of Richmond, Virginia; and that facility produced most of the iron for the cannons, shot, and ship plate utilized by the Confederate armed forces. With iron and coal discoveries in the west, iron mills appeared after the war throughout Tennessee and notably in Chattanooga; but it would be in Birmingham, Alabama, that a southern version of Pittsburgh would emerge.

Birmingham had been founded in 1871 by real estate developers who anticipated the city serving as a railroad hub. The city was ideally located to received the mineral wealth of the mountain areas to the northeast and quickly became a hub of iron production instead. The Tennessee Coal and Iron Railroad Company established the largest works and invested in new technologies to enter in a large-scale way into steel manufacture. In a move that was welcomed at first, the firm was purchased by the United States Steel Corporation in 1907. The Pittsburgh-based giant, however, soon developed pricing policies that placed Birmingham steel at a decided disadvantage, dampening developments there. Birmingham's steel industry suffered from the relatively low quality of the ores mined in the region and from limited markets in the South, but the decisions of northern capitalists took a toll as well.

Textile mills, mining, and iron and steel production provided notable signs of southern progress in manufacture in the postbellum period. There was also expansion in other industries. Flour and corn mills, tanneries, distilleries, and brickyards dotted the landscape. After the late 1870s, Winston-Salem and Durham, North Carolina, emerged as centers of tobacco processing and cigarette making; Memphis became the capital of cottonseed oil manufacture; a prosperous sugar beet industry appeared in Louisiana; and the port of New Orleans remained a busy place, with thousands of black and white longshoremen toiling at the docks. Most important, the forests of the South served as a prime source of timber. By 1900, in fact, the South supplied one-third of the lumber utilized in building construction throughout the country. The lumber of the region would also provide the base for a prosperous furniture industry that would emerge in the Piedmont after 1900.

Southern apostles of industry could point, then, to favorable developments. By the turn of the twentieth century, the South was home to a far greater number of manufacturing establishments than in 1860. Still, the accomplishments should not be overemphasized, for compared to the North, the South had hardly progressed. The region accounted for 30 percent of the nation's population but less than 10 percent of the country's industrial output. Most of the South's industries remained tied to the processing of agricultural goods; in fact, through the second decade of the twentieth century, lumbering remained the leading southern industry both in terms of employment and the value of goods produced. As to employment, while upwards of 60 percent of a rich farming state such as Ohio could be found in industry in the late nineteenth century, no more than 10 percent of the South's people labored in manufacture. The entire state of Georgia had fewer industrial workers than the city of Cincinnati alone. In overall economic terms, too, the South definitely lagged. By 1900, per capita income in the South stood at one-half that of the rest of the nation.

Problems in southern agriculture account for the region's poor overall performance, but why was there not greater industrial progress? A number of explanations can be offered. The South's late start in industry necessarily placed the region at a comparative disadvantage with the North. As Southerners often complained, the area also remained at the mercy of northern investors. Ninety percent of the stocks and bonds of southern railroads, for example, were held by Northerners, and decisions in the North could definitely jeopardize the South; the pricing policies of U.S. Steel provide a critical example. The South similarly lacked a base of technological knowledge and expertise and remained reliant on the importation of northern machinery and know-how.

The primary answer to the South's lack of industrial progress, however, lies in the area's low wage base. There was little pressure to substitute capital for labor, as noted earlier. Because of low wages, moreover, the region did not attract immigrants in large numbers, particularly immigrants with skills, who contributed so significantly to northern industrial development. Low wages also meant that purchasing power remained weak; there was not ample local demand to spur manufacture.

Finally, the elites of the South played a role here. The great landowners did not lose their power with defeat in war and the loss of their slaves; nor were they displaced by a new commercial elite; nor did they try to block initiatives in manufacture—points upheld by some historians. Landowners joined with merchants and professionals in sponsoring industrial growth, but together they moved cautiously, not wanting to disturb existing social arrangements or the place of agriculture in the society. Old and new men of influence shared in common a desire to avoid the formation of a rebellious working class like the one that seemed to have appeared in the North. They allocated relatively little public monies for education—keeping their taxes low in the process—and controlled industrial growth so as not to lure labor off of the land. The elites of the South also not only sustained existing racial practices, but oversaw a hardening of relations in the late nineteenth century with legislation legalizing segregation and the disenfranchisement of African-Americans. In keeping the white and black lower orders in place, they succeeded in keeping labor cheap and pliant and the South backward economically.

Change would eventually occur in the South. After World War I, blacks began to migrate to the North and the West; in the 1930s various federal farm and welfare policies either pushed or pulled labor (white and black) from the land. Tighter labor markets in the region brought mechanization in agriculture and a general shift toward capital-intensive means of production. Southern history became unfrozen, but only after an erosion in the low-wage labor base of the region.

The Geography and Dynamics of Late–Nineteenth-Century Industrialization

The great surge in manufacture after Civil War and the building of an extensive American industrial heartland (at least in the North) raise a number of larger questions. A first issue concerns location. Are there reasons why certain regions and places in the United States flourished with industry in the late nineteenth century? There are a number of obvious answers. To succeed, manufacturers needed easy access to raw materials and, in turn, markets; and cities graced with major waterways and railroads were at a clear advantage. Close proximity to energy reserves—water and coal—greatly benefitted development. Ready availability of labor—particularly skilled labor—and financial and marketing services similarly fostered success. Finally, there was a clear link between the prosperity of the hinterland and the fortunes of industrial enterprises in the late nineteenth century; most manufacturing firms thrived through the sale of products in local and regional markets, and where there was profitable agriculture, nearby industry benefitted in tandem.

Certain places had definite advantages, and the centering of American industry in the Northeast and the Midwest is not difficult to understand. But that is only one way of approaching and answering the question— to see locational choices as responses to given locational advantages. The history of industrial expansion in the late nineteenth century also divulges a proactive, and not just a reactive, aspect to the story. Chicago provides a telling example. The city had major physical liabilities, yet huge investments of energy and money made the site of this future metropolis suitable for expansive commerce and industry. Throughout the Midwest, there are countless examples of cities created in anticipation of (and not in response to) rural development and of promotional activity leading to the construction of transportation facilities.

Serendipity must be considered as a factor as well. There was no necessary reason for Trenton, New Jersey, to emerge as a center for wire cable manufacture, or for Cleveland to specialize in oil refining, and Detroit ultimately in automobile production. The presence of such individuals as John Roebling, John D. Rockefeller, and Henry Ford in particular communities made a difference. The greater role of elites in fostering, moderating, or blocking industrial growth is most expressively seen in the South. The location of industry, then, cannot be visualized as a function purely of resource endowments and market forces. Human will and social and political arrangements played their parts as well.

A survey of American manufacturing for the late nineteenth century also reveals different types of industrial cities: large and small, one-industry and diversified. (On the latter, it should be noted that there were no true single-industry towns; even in the great textile centers of New England, a variety of manufacturing concerns, from local machine shops to breweries, could be found operating in the shadows of the mills.) On the map of industry, there emerges no pattern in the location of the various kinds of industrial centers, although the largest cities appear to have been particularly well situated—able to tap the resources and markets of their immediate regions and well linked cross-regionally to other major cities. The major centers stood at the top of a definite urban hierarchy, New York and Chicago at the very apex. The primary cities had certain distinguishing characteristics. They were extensive production centers, they were key distribution points for the goods manufactured in satellite cities and the agricultural products harvested in the hinterlands, and they contained key financial and commercial institutions. In smaller urban locations business was not sufficiently substantial to sustain securities exchanges, merchandising marts, regional warehouses, and insurance and banking headquarters; and this explains the emergence of a hierarchy of cities in the late nineteenth century.

The contribution of America's smaller industrial centers, though, deserves emphasis. Attention can too easily focus on the Pittsburghs and Clevelands. One important aspect of the story of America's great leap in industrial production in the last decades of the nineteenth century is of the manufacture that occurred in the nation's seeming nooks and crannies (in the band of industry, of course, that covered the Northeast and the Midwest). Fine, diverse, and bountiful products issued from the shops and factories of the Zanesville, Ohios, and Grand Rapids, Michigans. These smaller places actually accounted for a significant portion of industrial employment in the era. In the Midwest, for example, 50 percent of all industrial workers resided in the region's smallest cities, and the other 50 percent, in the region's largest.

A shift did occur in terms of output. In 1860, the ten largest cities in the country accounted for 24 percent of the total value of industrial goods produced in the country; by 1900, that figure had risen to 38 percent. With industrial employment no greater in larger cities than in smaller, indications are that nation's mass-production-based firms were lodged primarily in the country's larger urban areas. There are standard economic reasons for the greater productivity of firms located in major urban areas; having access to greater supplies of raw materials, credit, and labor, they could purchase their resources on a greater wholesale basis and thus lower their total costs. But there is another element here, involving the discriminatory rate-setting practices of American railroads at the time. The railroads favored producers from larger industrial centers with lower shipping rates; freight cars leaving these cities were filled to capacity, and shippers of goods there received helpful discounts. Manufacturers from small towns were charged higher fees, which not only made them less competitive, but also less productive. The railroads (at least until the outlawing of discriminatory rate-fixing) contributed with normal market forces in this way to the particular location of mass-production enterprises.

Finally, in looking at the geography of industry, it should be noted that the history of America's industrial cities in the late nineteenth century is not entirely one of success. For some places, enterprise was remarkably short-lived. Troy, New York, for example, could be included in the top ranks of American manufacturing centers in 1870, but by 1890, the city was hardly recognized; the decline of its custom iron stove works spelled decline in Troy's total industrial fortunes. Similarly, as early as the 1880s, silk manufacturers in Paterson, New Jersey, whose efforts had just made Paterson America's "Silk City," were already looking to move their operations to eastern Pennsylvania to take advantage of the low-wage labor of women and children of coal mining families. Migration of textile firms from New England to the South commenced in similar fashion in the 1890s. "Deindustrialization" is an issue of concern for Americans in the late twentieth century, and while historians normally trace a first erosion in the country's industrial base to the 1920s, that history actually begins at the same moment as the building of the nation's manufacturing heartland.

In addition to questions of location, a survey of American industrial expansion in the late nineteenth century also raises questions of causation. The Civil War cannot be assigned a propelling role in the new burst of production. What did count then? Again, fairly straightforward answers can be provided. America's vast natural resource base greatly contributed to the more than fivefold leap in manufacturing output. There also seemed to be no end to the inventive (or at least, adaptive) spirit and genius of the American people. In 1860, 27,000 patents for new inventions were registered with U.S. patent authorities; by the turn of the century, the number of patents approved each year reached the one million mark. American inventions in the late nineteenth century continued to be based more on tinkering than theoretical understanding—there was little that passed for industrial research and development until the first decades of the twentieth century—but nonetheless, the new mechanical devices that issued from the nation's workshops contributed greatly to increased efficiencies and production. Some of the new machines were truly spectacular. James Bonsack devised a cigarette machine that increased the number of cigarettes manufactured each day from 3,000 to 120,000. Other notable machines dramatically changed soapmaking, grain processing, canning, distilling, and refining.

A finger then can be pointed to the supply side: ample supplies of resources and inventive talent guaranteed industrial success. There was an equally important demand side to the story. Great markets existed for the products of American industry, and here it was not just that transportation improvements allowed for a market, but that the vast growth of the American population in the late nineteenth century drove manufacture. And it was not just the absolute numbers of people, but the very composition of the population. While other modernizing nations at the time were losing rural populations, America was gaining with westward expansion of settlement and agriculture—that is, commercialized agriculture. While American farmers in the late nineteenth century faced extremely difficult circumstances, they still constituted a growing market for the products of American industry; and as discussed, the success of urban manufacture rested firmly on the development and relative prosperity of local hinterlands. Similarly of note were the swelling of urban populations and the growth of a strong urban consumer base for industrial products.

The great importance of population expansion to American manufacturing success underscores the importance of immigration. With fertility declines in the nineteenth century, immigration represented the prime means for population growth, and immigrants (whether they resided in the countryside or the city) served to boost demand for manufactured goods. American immigrants, of course, figured equally as much on the supply side as the demand, since they also provided labor for American industry. The nation had ample reserves of labor as well as material resources and technology for industrial progress. So critical is the factor of labor supply that economic historians estimate that 50 percent of the increase in manufacture in the late nineteenth century can be attributed purely to greater numbers of workers—in other words, more people just produced more products—and labor supply was twice as important a factor in growth as gains in either productivity or capitalization.

The immigrants' contribution in this regard was both quantitative and qualitative. Their numbers alone were important, for by 1900 more than 80 percent of America's industrial labor force were foreign-born workers and children of the foreign-born. There is an important demographic twist to this story because immigrants to the United States consisted disproportionately of males in their twenties and thirties, of prime age for heavy work. While a survey of American industry necessarily focuses on developments in the Northeast and the Midwest, not to be lost is the massive contribution of Asian immigrants on the west coast to railroad construction and mining. American industry was immigrant industry, and the nation's industrial workforce, which initially had been largely native-born, was re-formed through successive waves of immigration in the late nineteenth century. This had vast implications for American politics writ large and labor politics specifically.

The immigrants did not just add numbers; they added expertise as well. As a canvass of American industry after the Civil War indicates, immigrants continued to provide critical technical knowledge and ability for American manufacture, and whole industries owed their very inception and growth to the mass migration of particular groups of skilled European workers. Immigrants also added needed entrepreneurial talent and energy. In diversified manufacturing centers particularly, enterprising immigrants established the small-batch, custom manufactories that added so greatly to both the product mix of the country and overall production.

Immigrant enterprise raises the issue of entrepreneurship. From the historical record, it is clear that the nation did not suffer from a dearth of entrepreneurial spirit. There were great supplies of that as well (although distinctions could be made, and were during the time, between community-based, community-minded manufacturers and the grand acquisitors of the age). The subject of entrepreneurship leads to another angle on the causes of industrial growth in the late nineteenth century, and that is whether new techniques of management contributed to the expansion. The administration of American enterprises changed during the period, with the rise of the large-scale, bureaucratically managed corporation being the key development, and whether the progress of American industry can be attributed to new methods of business is an important question.

The question can be approached through a more general consideration: the extent to which developments in industry in the late nineteenth century represented a continuation of earlier initiatives or a fundamental shift. Can we dub what transpired in the period a second industrial revolution? There are changes to be noted, but the continuities appear greater than the disjunctures, particularly with regard to the actual production of goods.

Historians who point to an inherently new era of manufacture normally first emphasize the rise of capital goods industries—the so-called heavy industries. There is critical evidence to sustain this view. In 1860, for example, the four leading industries in terms of value of output were, in order, cotton textiles, lumber, boots and shoes, and grain milling. Three of the top four can be designated as consumer goods manufacture. By the turn of the twentieth century, the top four industries were, respectively, machinery, lumber, printing and publishing, and iron and steel. By then, three of the top four were producer goods industries. That machinery production claimed first place in American manufacture by 1900 is a clear indication of a shift in orientation. If distinctions are worthwhile, then America's first industrial revolution could be called a textile revolution, and the nation's second, a machinery revolution. There is an analogy in urban development: before 1860, American cities can be labelled as largely commercial, and after, as commercial and industrial.

There are grounds, however, to qualify any assertions of fundamental transformations in industrial structure. Stressing the emergence of capital goods industries loses sight of the impressive array of products that flowed from American manufactories. It is the *completeness* of the manufacturing system that deserves emphasis: Americans turned out steel rails and machines, but also clothing, ceramics, jewelry, and beer in great profusion. Visitors to the country in the late nineteenth century might be drawn to the imposing mass-production steel plants of Pittsburgh and the meat packinghouses of Chicago as they were to the textile mills of Lowell fifty years earlier, but missed again would be that large universe of firms in city and town that produced fine goods on a small-batch basis. Diversity and specialization persisted.

Continuities could similarly be viewed on the American shop floor. In small to medium-size firms, informality and personal relations still marked production. But even in new large-scale enterprises, old patterns endured and were only slowly undone and not without conflict. Skilled workmen supervised teams of men in iron mills, machine works, and mines and maintained great hold on the knowledge and pace of production; such practices were institutionalized where strong craft unions prevailed. In other instances, large firms operated on the basis of the so-called inside contract system. Executives divided their plants into separate shops and engaged men to run them as petty proprietorships. The contractors were paid by the piece and in some instances had to provide their own tools, labor, and materials. Major companies, such as Singer Sewing Machine, Winchester Arms, Remington, and Baldwin Locomotive, were managed for a time through inside contracting. Even in instances where firms hired salaried bureaucrats to supervise operations, arbitrariness reined. American railroads pioneered in the introduction of bureaucratic principles of management with clear lines of authority drawn (literally in the first flowcharts ever developed) and with very explicit rules and regulations to control all aspects of operations. As companies grew, despite an ordering from above, power devolved to local foremen and supervisors, who ruled in capricious ways. They exerted sway over recruitment, training, work assignments, compensation, fringe benefits, discipline, disposition in the case of accidents, and retirement, frequently favoring friends and relatives and others whom they could extort. A so-called foremen's empire existed not just on the railroads, but in many large-scale enterprises, and many strikes during the late nineteenth century involved grievances over the discriminatory rule of supervisors, with workers petitioning for union rules and regulations that would bring a modicum of justice and security to the workplace.

There were major attempts during the late nineteenth century to bring order to the American shop floor. Workers themselves campaigned for greater standards and procedures in the face of the capricious governance of foremen—and where successful they forced a kind of bureaucratization from below. Notable efforts were also made from above, but here two points must be emphasized. Managerial campaigns to exact greater controls at the workplace were battles fought on many fronts; the process was also ongoing, ever contested, and hardly complete (if ever completed) by the turn of the twentieth century. Management efforts to bring order to the shop floor included, among other strategies, the following: defeat of craft unions; mechanization to replace skilled workers; corporate consolidation of resources; ever more detailed divisions of labor; increased supervision and hiring of supervisors; the ending of inside contracting; the employment of unskilled and semiskilled immigrant laborers; the substituting of technical personnel and instrumentation for decision making by workers; the offering of various gratuities and benefits to workers to engender greater allegiances and discipline; the creation of career ladders within firms to foster loyalty; the perfection of machine tools, parts production, and standardized parts production techniques; and the adoption of conveyor-belt technologies to embed control of the through-processing of goods in machinery and not human hands.

Attempts at achieving greater and more detailed divisions of labor received particular attention during the late nineteenth century, and the figure of Frederick Winslow Taylor looms large here. Taylor, self-proclaimed originator of "scientific management" (or "Taylorism"), was born in Philadelphia to Quaker parents; instead of pursuing higher education as expected, Taylor became a machinist's apprentice and later a foreman at the Midvale Steel Company in his native city. At Midvale, Taylor began a series of experiments aimed at increasing the efficiency of the flow of goods through the productive process and the productivity of workers employed there; workers' control over the pace of production particularly aggrieved him. Although he introduced a range of managerial reforms, Taylor is most famous for his time-and-motion studies—his effort at breaking work into detailed, easily supervised tasks, cataloguing them, establishing expected rates for finishing jobs, and structuring incentive schemes to boost output.

Taylor moved on to serve as a consultant to many firms, and with his disciples and competitors he forged the scientific management movement. Taylor and others have been seen as critical agents in the changing of the American workplace, yet the historical record reveals that proponents of scientific management rarely succeeded in setting their innovations in place. Resistance from foremen who were threatened by the new consultants, more notable resistance from workers, and the administrative nightmare involved in cataloguing tasks and establishing rates—particularly in firms where product lines were always changing—doomed Tayloristic experiments from the start.

The actual history of scientific management qualifies the notion that the United States entered a distinctive new age of industrialism in the late nineteenth century. At the level of the shop floor and from the vantage point of most communities, the continuities with the past were notable. Even with regard to the famed American system of standardized parts production, great progress had not been made by 1900, despite new gains in metallurgy and machine technology. Production still slowed as skilled men honed and juggled with parts in assembling goods. Technical problems still remained. But beyond this, American consumers continued to be rooted in localized communities, buying locally made wares, and the demand and market for standardized products that would have forced greater improvements in standardized production techniques was still emerging, not emergent. On another front, the innovations that were to take place with electrification awaited the new century. Steam and flowing water still powered a majority of American manufactories. Finally, the growth of industrial production in the United States after the Civil War was based primarily on expanding supplies of labor, of more workers just producing more goods, an old pattern; greater output through improved productivity achieved by organizational and technological means awaited the twentieth century.

Industrialization had spread, America was now an industrial giant, and a refabrication of the American workplace had begun, but still, developments in the late nineteenth century at the lowest levels appeared to be more extensions of the past than an abrupt shift into the future. The last decades of the century, however, did witness the arrival on the American scene of a new presence: the large-scale, bureaucratically managed corporation. The big corporation, of course, contributed to the great industrial expansion of the era. While the impact of these new entities on the labor process was still in its formative stage, the corporation certainly represented and entailed news ways of conducting business. Understanding the origin and impact of the modern large-scale corporate enterprise is next in order.

Chapter Six

The Rise of Big Business

THE TYPICAL business firm of 1860 was small, family-owned and -operated (perhaps a partnership), specialized, labor-intensive, and a producer of small batches of goods sold in local and regional markets. The classic proprietorship persisted and proliferated during the late nineteenth century in small-town and metropolitan America and contributed greatly to the country's industrial success. By 1900, however, another kind of business concern took precedence. A different set of descriptives are needed to characterize these companies. They were large, corporately owned, bureaucratically managed, multifunctional, capital-intensive, and marketers of mass-produced items nationally and even internationally. The new firms also did not just blend easily with more traditional firms into the general fabric of American enterprise. They seized and occupied major terrain, garnering quick and widespread attention. Standard Oil, United States Steel, Armour, and American Tobacco were daunting companies whose names now surfaced daily in newspaper headlines and ordinary conversation. The men who figured imposingly in the establishment of the new behemoths of the age—the likes of John D. Rockefeller, Andrew Carnegie, Jay Gould, and J. P. Morgan—achieved even greater notoriety.

For some scholars, the emergence of the large-scale corporation in the last decades of the nineteenth century represents America's "second" industrial revolution. They view the corporation as part of a greater refashioning of American institutions at the turn of the twentieth century. But the equating of big business with a second industrial revolution creates a number of problems. First, different scholars emphasize different aspects of change in charting a supposed new course of development. For some it is the arrival of capital goods industries—with firms naturally organized as corporations because of the large-scale investments involved—that constitutes the essential departure from the past. For others, it is the adoption of mass-production techniques—again only affordable in the context of the corporation—that defines what is "new." For still others, transformations in the organization of business that are entailed in corporate management mark one age from another. Disagreement and confusion over the exact nature of change renders the notion of a revolution problematic.

The term *revolution* provides other traps. To stress the inception of producer goods industries, for example, misses the very comprehensive growth of industry in the late nineteenth century and the sustaining of America's diversified manufacturing system. Similarly, to pinpoint new automated production techniques overstates the extent of technological change and neglects the persistence of old practices and ongoing conflict on the shop floor. Highlighting the organizational innovations and breakthroughs of corporate management, likewise, disregards the world of business that continued to transpire outside the corporate realm in traditional enterprises; but more important, it fails to convey the actuality of administrative change as a slow and staggered process.

This chapter will focus on the origins of the corporation, treating different ways of understanding the rise of big business in the late nineteenth century. The new corporations of the age can be seen respectively as the handiwork of extraordinary men, the outcome of particular events and circumstances of the period, or the product of impersonal, long-range demographic, economic and technological forces.

Commentators of the day had little trouble accounting for the rise of the new mass production firms. The large-scale corporations were simply the grand inventions of their founders, and depending on the politics or viewpoints of the reporter, the founders were either afforded great praise or blame. Scholars, later, also placed the creators of these enterprises on center stage, and again depending on the interpretation of the individual writer, the founders were referred to as either remarkable innovators or devious robber barons. In recent times, historians have shifted attention away from the notable business figures of the age toward a different set of actors: namely, the managers of the new firms, the men appointed to devise organizational schemes, coordination mechanisms, and production and marketing strategies that allowed the new behemoths, once born, to survive and thrive. This important new way of looking at the corporations does share with earlier approaches a stress on the role of human agency in historical change.

Another kind of shift also marks recent scholarship: attributing developments to specific contemporary events or to long-range impersonal forces. Occurrences and situations in the late nineteenth century that permitted the formation of big businesses, such as key legal decisions and the economic depressions of the period, are thus cited. The new companies are also treated as the results of the simple expansion of population and market activity or of various technological imperatives.

A discussion of all possible explanations—with some weighing of alternative approaches—can provide for a fuller understanding. Each view will be discussed to afford a greater handle on developments, but also to be emphasized is that the new corporations were not fixed entities. Many of these famed enterprises failed; others succeeded, but only after numerous readjustments. The new corporations faced both pressure from the outside and conflict from within, and trial and error marked their histories.

The Founders

A fascinating cast of characters appeared seemingly out of nowhere in the United States during the late nineteenth century to lead what can ostensibly be termed a "revolution" in business techniques. The men involved conceived of enterprises that had no precedents in either size or complexity. They moved audaciously in the business and political realm, cornering markets and politicians at will. They also amassed incomprehensible fortunes. The great business figures of the age can be lumped together because they shared traits in common. Yet, they acted in rather disparate ways, and the differences deserve articulation. If they are examined as a group, several categories emerge: they were speculators, empire builders, integrators, promoters, and inventors.

Speculators

One set of businessmen accumulated their great wealth through the frequent buying and selling of companies (railroads were their pet projects) and the related manipulation of the securities market. A number of such wheelers and dealers can be mentioned—Jim Fisk, Daniel Drew, and Jay Cooke were among the great players—but the undisputed master of the speculative scheme at the time was the infamous Jay Gould.

A story is told of Jay Gould when he was a young man that summarizes his entire life. Gould grew up on a farm in upstate New York in the 1840s and as teenager worked as a clerk in a village store. One day he apparently overheard the owner of the store negotiating for the purchase of land and offering $2,000. Gould then quickly borrowed $2,500 from his father to buy the property. He subsequently sold it to his employer for $4,000, netting a handsome profit. There is some justice in this tale, for Gould would soon be dismissed from his job. Gould later in life would be similarly censured (though never really punished) for his extraordinary and continuous violations of people's and even the public's trust, but he never ceased to walk away from his schemes with money bulging in his pockets.

With cash from this and other early ventures, Gould in his early twenties next became part owner of a large tannery. He soon began using the assets of the company for forays into the stock market. He personally did well, but his speculations led to the insolvency of the business and the suicide of his partner. During the Civil War, Gould bought and sold government bonds, gold, and currency to great advantage; and after the war he used his profits to move heavily into the purchasing and selling of railroads. He would typically buy small lines that major carriers could use to complete their networks; he would then hold out, encourage rumors of imminent sales to boost the value of his stock holdings, and finally sell his shares at enormous profit. Other tactics included the creation on paper of a new railroad company, the threat to build his railroad parallel to a competing line, and then the sale of his never-built road to the threatened parties. Gould also made a practice of issuing new, so-called watered stock to himself, which artificially boosted the value of the assets he had to sell.

With Jim Fisk and Daniel Drew, Gould also purchased a commanding number of shares of the Erie Railroad and this brought him to an epic confrontation in the late 1860s with Cornelius Vanderbilt, the great shipping line magnate and then founder and owner of the New York Central Railroad, a direct competitor of the Erie. Vanderbilt tried to purchase shares of the Erie to control the company's management, but Gould and his allies foiled Vanderbilt by illegally flooding the securities market with newly printed Erie stock. The three Erie conspirators then fled to New Jersey when New York State authorities ordered a stop to their activities. Advancing lavish bribes to state assemblymen in New Jersey, Gould secured passage of legislation that legalized his continuing hold on the Erie Railroad. Vanderbilt was forced to retreat.

Gould never actually attended to the management of the Erie—his wealth was always based not on the successful operation of his enterprises, but on his ability to inflate their value and then sell off—and soon the carrier would experience severe financial difficulties. Gould schemed to add new railroads to the line, expanding Erie train service from New York to Chicago; but soon, other investors in the company challenged his leadership, and he was forced out of management (though, as usual, without personal losses).

Meanwhile, Gould engaged in another speculative venture that proved to have deleterious effects for the entire country. In 1869 he hatched a plot to corner the nation's supply of gold. With gold prices low at first, he pooled his and his associates' resources for the massive buying of gold. He then leaked rumors that he was acting as an official representative of the U.S. government in purchasing gold, which immediately drove up the price of the precious metal; he also tried to prevail on his contacts in Washington to guarantee that the government would not release any of its holdings of gold into the market to stem the speculative mania. Eventually, the president of the United States, Ulysses S. Grant, was forced to disavow any governmental connection with Gould, and he ordered the Treasury Department to start selling gold. Prices then plummeted as investors tried to unload their holdings with the least losses. With many now ruined, general investment activity was curtailed. Gould, of course, had sold all the gold he had purchased before the president's announcement, at peak prices. He was roundly condemned for jeopardizing the nation's economy, but once again he walked away from the damage done enormously enriched.

Gould never relaxed. For the next twenty years until his death in 1892, he was engaged unceasingly in speculative ventures. What is remarkable about his life is not only the daring and immensity of his schemes, but the long time span of his activities and his relentlessness. In the 1870s he moved to gain control of the Union Pacific Railroad and through the buying of lesser lines tried to create a truly transcontinental railroad. This enterprise never came to full fruition, and he pulled out before others were left to pick up the pieces. In the 1880s he further speculated with railroads in the Southwest and successfully forced the Western Union Telegraph Company to purchase a rival communications firm that he had begun to fashion. He would wind up with a controlling interest in Western Union. In these ventures, as before, he never took part in the actual management of the firms he bought; they existed as items to be sold when the price was right.

Jay Gould had his associates. There was Jim Fisk, who made his first fortune selling contraband cotton during the Civil War and who would hook up with Gould in his speculative railroad and gold ventures of the late 1860s. There was also Daniel Drew, who made money as a young man buying and selling cattle; later he became a successful stockbroker, banker, and co-investor with Gould and Fisk.

Gould also had his competitors, of whom Jay Cooke was among the most formidable. As a young man, Cooke first secured through family connections a job in a shipping company; he later joined a banking house and became a successful bond salesmen. In 1861 he formed a banking firm of his own (Anthony Drexel would become a partner), and again family connections helped as he became a chief agent of the U.S. government for the marketing of government securities during the Civil War. So expert was Cooke in selling war bonds that with $2 million worth of bonds sold a day at a 0.5 percent commission, he was earning $3 million a year for his services to the government. With his great profits, Cooke turned after the war to speculating in railroad development in the central and Pacific Northwest. A failure of one of his ventures led to a complete collapse of the stock market in 1873 and was a key cause of a seven-year-long general economic depression that then ensued. Cooke faced off directly with Gould in Gould's attempt to corner the gold market in 1869. Jay Gould would later have other competitors like Cooke, particularly in railroad schemes—the names of Henry Villard and Edward Harriman come to mind—but in business speculation in the late nineteenth century, he had no match.

Empire Builders

There was another set of leading business figures of the age who were also not above speculative activity, but they made their fortunes in overseeing the building of single, mammoth, and enduring enterprises. Two of the most famous businessmen of the period, Andrew Carnegie and John D. Rockefeller, fit in this category.

Andrew Carnegie was born in Scotland. His father was a handloom weaver who became active in radical working-class politics in Scotland. Of all the major businessmen of the late nineteenth century, Carnegie was the only one to express grave misgivings about the big corporations and their threat to democratic and republican values and institutions, and his heritage played a role here. Living in desperate circumstances, the Carnegie family immigrated to the United States in 1848, settling in Pittsburgh, and the young Carnegie occupied a number of low-paying jobs until he secured a position as telegraph operator on the Pennsylvania Railroad.

To Carnegie's good fortune, he came under the wing and tutelage of Thomas Scott, superintendent of the Pennsylvania, and later the carrier's forceful president. Scott not only taught Carnegie important managerial skills, but he also tendered him advice on investing in the stock market. Scott took Carnegie to Washington during the Civil War when he was appointed to coordinate rail traffic on behalf of the Union cause, and the young Carnegie received additional lessons on the workings of government. By war's end, Carnegie had risen to a divisional superintendency on the Pennsylvania and had also accumulated a small fortune from his investments.

Carnegie chose not to stay with the railroad. Instead, he assumed management of an iron bridge construction company in which he had invested during the war and attended more fully to speculating in the stock market. From his work in railroads and bridge construction, Carnegie became convinced that great opportunities lay ahead in iron and steel production. In the early 1870s he therefore turned all his energies toward the building of a major iron and steel mill with the latest technologies, including Bessemer convertors. In 1875, he opened the Edgar Thomson Works (named after the first president of the Pennsylvania Railroad), by far the most modern facility in the country. Carnegie innovated not only with new machinery, but also with new administrative and accounting techniques that allowed him in three years to achieve a 31 percent return on his original equity. Over the next decade he built and bought additional facilities, making the Carnegie Steel Company the major firm in the trade.

Carnegie succeeded by adopting new technologies—he eventually replaced his Bessemer convertors with even more efficient open-hearth furnaces—and also by hiring managers who perfected production and accounting systems that minimized costs. Carnegie himself began to spend less and less time near his plants. To help with management, he brought Henry Clay Frick into the company. Frick owned massive coke fields in western Pennsylvania and operated huge coke smelters. In bringing Frick into his firm as a partner, Carnegie not only gained control over a supply of the key fuel for steel production, but he also gained an extremely able general administrator for his company. Frick assumed management of Carnegie Steel and oversaw its further expansion, modernization, and integration; under his leadership the firm freed itself from outside suppliers of goods and services by purchasing its own iron ore reserves and establishing its own transportation facilities.

Tensions grew between Carnegie and Frick in the 1890s, however. Frick's brutal handling of the 1892 Homestead Strike was one source of contention. Carnegie slowly became amenable to the idea of selling his 60 percent share of the giant firm. In 1901, the Carnegie Steel Company was thus absorbed into the new United States Steel Corporation, a monumental conglomerate glued together by the financier J. P. Morgan. Carnegie received $250 million from the deal, and with this money he began his second career as one of the world's most prominent philanthropists.

John D. Rockefeller rivalled Andrew Carnegie as a great donor to American cultural, educational, scientific, and religious institutions, yet his motives were always under suspicion. While Carnegie's philanthropy represented an attempt to reconcile his formidable wealth and power with the egalitarian values he learned as a child, Rockefeller's charity appeared to be more a public relations ploy than the result of deep-seated ethical convictions. Rockefeller oversaw the assembling of an industrial enterprise dwarfing even Carnegie's; but the manner in which he conducted business made his name synonymous with ruthlessness.

Rockefeller grew up in upstate New York. As a boy he exhibited many of the traits for which he would be famous: extreme piety, diligence, and frugality. He raised and sold turkeys and hoed potatoes long hours of the day, accumulating sufficient savings actually to advance loans to the farmers who employed him. Rockefeller moved with his family to Cleveland, where he learned bookkeeping skills in school and as a clerk for a produce merchant. With $800 in savings, he opened his own business in 1859 and prospered in the buying and selling of grain and the provision of goods and credit to his rural customers. The Civil War proved a boon to the young Rockefeller; with rising prices, Rockefeller earned more than $17,000 a year marketing grain and other agricultural products to the Union army and in northern urban areas where commodities were scarce.

Rockefeller initially invested his surplus capital in railroad stocks and land, but in 1863 he was attracted to another opportunity. After scouting the new and burgeoning oil fields of northwestern Pennsylvania, he decided to join with some business friends in sponsoring the construction of an oil refinery in Cleveland. Successful in this venture, Rockefeller at war's end closed his store to concentrate solely in the oil refining business. He poured his profits into the building of the nation's largest refinery, which by 1867 was operating at a capacity of 1,500 barrels a day, more than twice that of his nearest competitors.

His competition bedeviled him—in this nascent industry, new oil wells and refineries sprouted daily—and Rockefeller determined that profits were to be made not just in continued technological and organizational improvements, but in curtailing, if not utterly removing, competition. At first, he chose simply to buy out his rivals. With the help of a few partners and leading Cleveland bankers, he was able to pool monies to purchase fifty competing firms in Cleveland and eighty in Pittsburgh alone in the years 1865 to 1868. But Rockefeller was soon to opt for two other tactics that would lead to the building of his domineering Standard Oil Company.

First, he reached secret agreements with various railroad carriers to have his oil shipped to eastern cities at substantially lowered rates. Fierce competition prevailed among eastern carriers at the time, and since Rockefeller could guarantee large shipments, they fought and bargained for his favor. Rockefeller's refined oil products—kerosene for lighting, most notably—thus sold at cheaper prices in distant markets, effectively damaging his competition. Rockefeller's efforts were not without opposition. A collective scheme between Rockefeller and other major refiners, involving not only lower transportation rates but also rebates to the refiners (who received a portion of the elevated fares charged the competition), led to complaints from oil drillers and independent refiners and accusations of conspiracy in the press; the uproar forced state legislators to be less lenient in the corporate charters issued to Rockefeller and his group. But while Rockefeller's reputation suffered from the controversy surrounding his favored treatment by the railroads, his business did not. He continued to receive advantageous rates, and he moved to build his own transportation facilities as well.

Now, however, Rockefeller chose another means to curb what he saw as ruinous competition in the refining business. This was to invite his rivals into a new conglomerate operation. After 1873 Rockefeller began to offer other refiners cash or newly minted Standard Oil stock in exchange for their coming under the wing of his company. He employed all forms of persuasion to convince his competitors, and by 1878, the firms who had joined with him accounted for more than 80 percent of the refinery capacity in the nation. The group could control the price at which crude oil was bought, the cost of transporting refined products to market centers, and their price to dealers and customers.

The governance of this empire proved a source of difficulty. Technically, the firms participating with Rockefeller remained separate entities; they did not necessarily have to abide by the orders of a board of directors established to coordinate activities. Legally, too, the Standard Oil Company, which was incorporated in Ohio, could not hold the assets or stocks of other companies, particularly those chartered in other states. In 1881, Rockefeller's lawyers happened on a possible solution. A board of trustees was created to hold all stock of subsidiary firms and Standard Oil securities; they would, in turn, issue new certificates of inclusion in what was deemed a trust estate. The Standard Oil Trust and the whole idea of a business "trust" was thus born. The trustees, with Rockefeller on top, became the formal directors of the empire, making decisions for all to abide by (although again in practice, the prerogatives of the participants remained hazy).

The Standard Oil Trust came under immediate attack, not only because of the trust's economic power and conspiratorial appearance, but also because it was unclear whether individual states could tax the properties of the amorphous giant. Public sentiment grew for legislation to outlaw combinations that monopolized and restrained trade, and in 1890, with the specter of Rockefeller's Standard Oil Trust hovering over its proceedings, Congress with almost unanimous approval passed the Sherman Antitrust Act. In 1892 state officials in Ohio also successfully brought suit against Standard Oil on the grounds that it violated its state charter in carrying on business outside the state. Rockefeller quickly dissolved the trust and reorganized his business into several companies chartered in different states. With interlocking directorships, a key set of executives maintained control over the whole. In 1899, new laws in New Jersey allowed Rockefeller to place authority for the several Standard Oil companies into one holding company located in that state.

Despite all these legal entanglements and changes, Rockefeller and a small group of associates had effectively retained power over all the subsidiaries of his empire since the 1870s, when he had decided that financial success rested with the controlling of competition through combination. And successful he was. In the 1890s, Rockefeller's Standard Oil properties were earning $45 million a year in profits. After donating hundreds of millions of dollars to charity, his personal wealth was still placed in 1913 at about $325 million, and it was estimated that he had earned up until that point in his life a total of at least $1 billion.

There are a number of ways to account for Rockefeller's immense fortune. He dominated a new industry almost from its beginnings with great business acumen. He innovated with the newest technologies, and his refineries operated to the peak of efficiency (of course, he was dealing with a product and a process that lent itself well to mass production and the achievement of economies of scale). He appointed an administrative staff who developed effective flowcharts, coordination procedures, and accounting techniques. Rockefeller knew how to cut costs—an obsession with him—because he knew his costs at each stage in the production and marketing process. He wisely expanded his operations beyond refining oil to control the accessing of needed raw materials, their transport, and then the transport and sale of products. He also quickly built Standard Oil into an international operation. And he kept a tight hold on the various pieces of his empire, even as the means to hold the firms he incorporated under his umbrella changed. Thus, we can attribute the rise of Standard Oil to Rockefeller's administrative foresight and ability. But equally important was his basic strategy: to seize a commanding position within the industry by conquering and absorbing his competitors. Rockefeller used any and all means to build his empire: secret pacts and coercive threats and moves against his competitors that bordered on the illegal; armed responses to workers who protested conditions of work in Rockefeller enterprises. He earned a reputation for ruthlessness, and with justification.

Big businesses emerged in the late nineteenth century not just because speculators made money off their formation and sale, but because another group of business leaders devoted their constant energies to the creation of singular concerns. To be mentioned also among the empire builders are the likes of Cornelius Vanderbilt, who established one of the country's largest shipping lines and then oversaw the building of the New York Central Railroad; and Collis Huntington and James Hill, who assiduously assembled major western railroads. But when the conglomerated giants of the age are recalled, Andrew Carnegie and certainly John D. Rockefeller draw our immediate attention.

Integrators

Another set of businessmen of the late nineteenth century also crafted single giant firms, but not primarily through agglomeration and merger. Rather, they took businesses and expanded them into multidimensional enterprises. Their innovations in management are striking, and separate treatment is in order. Gustavus Swift provides probably the best example of this other kind of empire builder.

The die was cast for Gustavus Swift when he was fourteen years old and living in a small town in Massachusetts. He then went to work for his brother, the village butcher. Two years later Swift ventured out on his own, buying a heifer, slaughtering it himself, and then peddling the cut meat door-to-door. By his early twenties, Swift had become a leading cattle dealer, butcher, and salesman of dressed beef in the region. His sights were set higher, though, for Swift realized that the center of the opening national market for packed meats would be in Chicago, and in 1875 he moved his business there.

Modernized slaughterhouses had already been established in Chicago by the time of Swift's arrival, but live cattle were still being shipped to eastern markets to be sold and butchered there. This was not a profitable enterprise: animals died en route or else lost great bulk; the shippers' transportation costs were also based on the original weight of the animals, and ultimately only 40 percent of their mass was edible and butchered and sold. Swift wanted to take advantage of the low cost of cattle in the Chicago market and of mass-production techniques in the city's slaughterhouses. Shipping and selling cut meats appeared to be the goal; the great problem was how to preserve the meat in transport. Swift initially decided to send dressed beef by rail to eastern markets only during the winter months (in open boxcars along the northernmost routes). To achieve the greatest economies, however, he needed to operate on a year-round basis, and so he quickly turned to another notion: that of refrigerated railcars. Working with an engineer, Swift designed a car in which ice was packed under the roof and kept the meats in the car cool through convection currents. In 1881 he shipped his first railcar of dressed beef to cities in the East.

Had Swift done nothing more than develop the refrigerated boxcar, he would have entered the annals of great business achievers, but he went much further. He had safely shipped his dressed meats hundreds of miles, but now two new problems emerged, the first of which was what to do with the meats on their arrival in eastern cities. Swift solved this issue by creating a network of refrigerated warehouses. More important was a problem of sales: how to convince local butchers to market and customers to purchase meats slaughtered in a far distant place many days earlier. Here, Swift created a pioneer sales force. Locally based salesmen assumed the responsibilities for inducing butchers and storekeepers to handle Swift products, for public advertising, for coordinating orders from warehouses to stores, and for keeping executives in Chicago informed of local demand so that production and sales could be coordinated. Swift ultimately built a nationwide and international sales force; by the turn of the twentieth century his company was, in fact, marketing dressed meats overseas.

Swift's business expanded by great bounds after 1880—the firm had more than twenty thousand employees in 1900—but not through combination and the absorption of the competition. Rather, his company grew through the assumption of new activities. Swift created separate divisions to handle the purchase of livestock, their mass slaughtering and processing, the transport of products, their warehousing, and their marketing. Administrative offices in Chicago oversaw the operations of the whole and devised plans for expansion. While Carnegie and Rockefeller also fashioned multidimensional and integrated firms, Swift built his empire in a distinctly different way, and his innovations are noteworthy.

There were also other businessmen who created large enterprises in the late nineteenth century primarily through expansion of the functions of their firms and not the engorgement of their rivals. Philip Armour built a meatpacking firm to rival Swift's and in the same integrated fashion. John Dorrance invented a process for canning condensed soups and founded the Campbell Soup Company. His firm grew through the development of different product lines and great attention to marketing: American homemakers had to be convinced not only to buy foods they normally prepared fresh for their families, but also to buy them in newfangled cans. Theodore Vail took Alexander Graham Bell's invention of the telephone and constructed the multifunctional and multidivisional Bell Telephone Company (later American Telephone and Telegraph).

James Duke can be added to this list of business integrators, but with qualification. Duke, one of the few great business figures of the period born in the South, assumed direction of his family's tobacco company at the early age of twenty-five. Duke decided to concentrate on the production of cigarettes, a relatively new tobacco product in the marketplace. In addition, in 1884 he purchased two of the cigarette machines developed by James Bonsack which allowed him immediately to increase tenfold the daily production of cigarettes. Gaining sole rights to the machine, Duke now easily outproduced his competitors, but he faced a new problem: the saturation of what remained a small market for cigarettes.

Determined to create new customers, Duke employed a large sales force; his innovations in sales, particularly in advertising, would be as pathbreaking as Swift's. He also moved to consolidate the buying and processing of tobacco leaf to maintain production at full capacity as sales did indeed mushroom. Success was instantaneous. In 1890, Duke's company earned $400,000 in profits on the $4.5 million sale of 834 million cigarettes. As of that year his firm stood as a prime example of the building of a business empire through the novel expansion of functions. However, in the same year Duke adopted a different strategy. Competition in the industry was now an obstacle to continued high earnings, and Duke and four of his competitors merged together to form the immense American Tobacco Company. As a result of this move, Duke's story resembles Rockefeller's more than Swift's. Duke, like Rockefeller, would emerge as one of the richest of the new captains of industry; his conglomerate enterprise would also face similar publicity and legal problems.

Promoters

The businessmen considered to this point played direct roles in the rise of large-scale enterprise in the United States in the last decades of the nineteenth century. They built the awesome companies, whether purely for speculative purposes or to generate income from the sustained production of goods. Another set of actors also had an essential part in the emergence of big business, but they operated in the background, and their impact was indirect. Investment bankers comprised this group, and they affected developments by providing critical services to the company builders: they loaned them money; they underwrote and sold stocks and bonds for them to raise additional capital; they helped the insolvent by reorganizing and refinancing their businesses; and they encouraged and facilitated mergers of firms, marketing too the new securities of the conglomerated concerns—all this assistance rendered, to be sure, for most handsome fees. The bankers' contribution to the rise of big business was sizable: that there was great money to be made simply in the promotion of large enterprise is one way of accounting for the new behemoths. The investment bankers occupied an important place, and one among them wielded extraordinary power and influence: John Pierpont Morgan.

J. P. Morgan was born a finance capitalist. His grandfather and father had amassed fortunes in insurance, commission merchandising, and stock brokering. In 1854, his father, Junius, had become a partner in a powerful international banking firm based in London which eventually became J. S. Morgan and Company. The firm dominated the marketing of securities of American companies, mainly railroads, in Europe and acted as the agent for a number of European governments in the securing of loans. At the age of nineteen, J. P. Morgan entered his father's business, serving as his representative in New York City. The Morgan family's contacts in Europe would be a major asset in the younger Morgan's later success in finance.

In 1861, J. P. Morgan founded his own firm in New York, J. P. Morgan and Company, which was an official branch of his father's bank, but soon he began to chart an independent course. During the Civil War he made money speculating in gold. After the war he helped defeat Jay Gould in a few of his efforts to control railroads, and with the collapse of Jay Cooke's banking house in 1873, Morgan became the principal agent of the U.S. government for the marketing of treasury bonds.

In the 1880s, Morgan began to make his real mark. European investors were becoming leery of the speculative railroad ventures of Gould and others, and Morgan moved to restore faith in American securities. He started holding meetings with railroad executives to convince them to abandon their predatory schemes and curtail competition. He settled a number of disputes between major carriers, forestalling the construction of various parallel road projects, and encouraged the rail lines to reach agreements on the pooling of traffic. During a major depression starting in 1893, Morgan was called upon to reorganize and refinance several major bankrupt railroads, and in collecting his fees in new stock issues, Morgan found himself and his associates sitting on the boards of directors of many lines. He could now enforce his cooperative agenda: the railroads would work with and not against each other to achieve rationalized service and stable profits.

During the depression of the 1890s, Morgan, a relatively unknown figure outside business circles, also found himself for the first time the object of popular attention and derision. President Grover Cleveland had appointed Morgan as the government's chief agent to purchase gold for the Treasury. Morgan earned a sizable commission for his assistance, but Cleveland's tight, gold-based monetary policies exacerbated an already spiralling deflationary situation and brought no relief to the great majority of Americans, whose hardships were extreme. Morgan became a perfect cartoon figure, the fat banker or industrialist of the age who made millions while the many suffered.

Morgan was undeterred by popular attacks. During the late 1890s and in the first decade of the twentieth century, he would orchestrate several extraordinary mergers that would see the formation of such corporate giants as General Electric, United States Steel, and International Harvester. Morgan's place in economic affairs during the period was unusual. He was not just a banker or a promoter. He was an overall systematizer as well. He believed that by working with each other, the major corporations could bring economic prosperity and stability. He would even show his fellow businessmen that corporate leaders could work with top government officials to effect greater order. Morgan represented a vision of a new, organized America.

Morgan had no rivals. The Mellon family of Pittsburgh were key bankers who helped finance Andrew Carnegie, major railcar construction firms, oil businesses, and aluminum producers; Wall Street was also home to other powerful investment houses. But no one matched Morgan in his attempts to mediate within trades, in the magnitude of the mergers he facilitated, and in his macroscopic view.

Inventors

The businessmen who founded the new large-scale enterprises of the late nineteenth century generally latched onto particular products or services which they determined could make their fortunes; they had no necessary prior knowledge or expertise with the product or service. A small group of men who played a role in the rise of big business in the period etched out their places through the actual invention of new products and then the creation of businesses for their manufacture and sale. George Westinghouse and Thomas Alva Edison are prototypes here.

George Westinghouse received his first patent at the age of twenty-one in 1865 for a rotary steam engine he perfected. He was employed at the time in his father's farm implement factory. Just four years later, though, he would invent a device that would make him world-famous, an air brake for trains. (Before the general adoption of Westinghouse's automated brakes, trains were slowed and stopped with hand-turned brakes.) Westinghouse established the Westinghouse Air Brake Company to manufacture braking systems, and his firm prospered as railroad companies chose to install the new, safer brakes or were forced to by government regulations. Westinghouse also began to experiment with electrical railroad switches and signals, and in 1882 he established another company for their manufacture. The 1880s were a flourishing time for him; during the decade he received no fewer than 134 patents for his additional inventions, including various electrical apparatus, gas control devices, steam turbines, and railroad equipment.

Like other inventors of the time, Westinghouse began to turn all his attention to the new, expanding field of electrical generation and use. In 1885 he founded the Westinghouse Electric Company to manufacture electrical generators and appliances. He gambled on producing goods that were based on alternating current (ac) and this would bring him into direct conflict with Thomas Edison and a group of investors associated with Edison and his direct current (dc) electrical system. Westinghouse would win the ensuing scientific war but lose on the business front. Alternating current could be boosted through transformers to high voltages and then transmitted over long distances with little loss in power; it was also more suitable for newly developed electrical motors. By the early 1890s, it was clear that ac was the wave of the future; the only advantage of low-voltage dc was its relative safety, which Edison featured in his appeals to customers. Edison, however, had the backing of the banking house of J. P. Morgan. In 1892, Edison's firm, Edison General Electric, merged with Thomson-Houston Electric, a major independent in the field, to create the General Electric Company. GE not only had the resources to outcompete Westinghouse in the marketplace in delivering both ac and dc electrical generators and appliances, but also the financial wherewithal to fight Westinghouse in the courts over patent rights and infringements. The two companies eventually settled on a pooling arrangement of patents, but greatly to Westinghouse's disadvantage. While many of Westinghouse's businesses continued to flourish, even internationally, his electric supply firm suffered financially in battle with GE and entered bankruptcy. He was ultimately bailed out by a group of investors who assumed management of the company. Westinghouse would not live to see his Westinghouse Electric survive and grow to be second only, but second by far, to GE in the electrical goods industry.

Thomas Edison is not readily associated with the world of business; he is the great figure of invention of the late nineteenth century. Edison, in fact, was involved with any number of business ventures connected to his inventions, although in contrast to Westinghouse, he was more business initiator and pitchman than developer and manager.

Edison was born in Ohio in 1847. After showing little promise in school, he secured a job in his late teens as a telegraph operator. He began tinkering with the telegraph equipment and devised an improved stock ticker. With money raised from the sale of the rights to his invention, Edison opened a business to continue his technical explorations. In the mid-1870s, he and a team of employees worked on new telegraphs; they were only slightly behind Alexander Graham Bell in perfecting a telephone; in 1877, they produced the first phonograph, perhaps Edison's only sole invention; and they made great improvements in incandescent lights (Edison did not invent the incandescent bulb). Edison's own business, it should be noted, was also something of an invention, for he fashioned a pioneer industrial research laboratory.

In the early 1880s, Edison further led the way in electrification. He designed the nation's first electrical power station (in New York City), oversaw the installation of wiring and lighting and other equipment in homes and businesses, and created a pathbreaking system of electrical service. In characteristic fashion, he formed a company to provide electricity and appliances, designed and organized the complex technology, promoted consumership for the new product, and then summarily left the management of the operation to others as he moved to other experiments. The Edison General Electric Company was his creation, but by the time it became a foundation block of the giant General Electric Company, he had little association with the actual business. He had a similar involvement with the motion picture industry. His laboratory perfected (but he did not invent) the basic equipment for moving pictures; Edison then went on to organize and boost the industry, but he would play no direct role in the management of firms within the trade. Later in life the pattern would be repeated with his inventive work in electric storage batteries, railway signals, train and mine lighting, dictating and mimeograph machines, submarines, iron refining, and even cement manufacture.

Thomas Alva Edison had a hand in the emergence of big business in the United States in the late nineteenth century, but his impact was as intangible as it was concrete. He invented and perfected numerous products and founded what became large companies to manufacture and market them, as did his fellow great inventors of the age, men like George Westinghouse and George Eastman (who did attend to the management of their concerns). Edison's influence, however, extended far beyond his inventions and the firms with which he was associated. He played a role as salesman for science and the new technologies; as conceiver and initiator of system-based enterprises; and as a symbol of modern American innovation and progress.

In emphasizing the role of businessmen in the rise of large-scale enterprise in the late nineteenth century—which is only one way of accounting for developments—the range of players has to be noted: the Jay

Goulds as well as the Gustavus Swifts and Thomas Edisons. The varied nature of entrepreneurial activity even on the single high plane of big business must be appreciated. Historians, however, have stressed the commonalities of the business moguls of the period. When biographical information for a large cohort of them is assembled and analyzed, they do appear to be cut from the same cloth. They were all white and male, for starters. Upwards of 90 percent were also Protestant, and 75 percent of Anglo-Saxon heritage. The great majority grew up in New England and in urban environs. While there are a number of rags-to-riches tales, over 85 percent of the business leaders of the age of corporate growth were sons of businessmen and professionals. They embarked in enterprise with great advantages. As the representative biographies above indicate, many also began their careers in business as clerks in merchandise operations. They received early experience in the buying of goods cheap and the selling of them dear. The speculators later extended on their apprenticeships: they bought and sold firms. For those who moved from commerce to industry, their schooling in marketing would serve them well. Finally, the great business figures as a group were born in the 1830s and 1840s, and all escaped enlistment in the Civil War; they either bought themselves out of duty or paid for substitutes, as was allowable at the time. They were thus the lucky ones of their generation. They were also fortunate to be in positions to prosper during the war in providing goods and services for the war effort.

Thus, the leading business figures of the late nineteenth century had common backgrounds and early experiences. But, as we have seen, they did participate in the building of large-scale enterprises in varied ways. Their later professional and personal lives were also filled with contradictions which makes a generalized assessment difficult. For example, these were men who with few exceptions were quite pious, frugal in business, and dour in demeanor; yet with their lavish spending on mansions, social events, world tours, and art, they rightfully became identified with gross materialism and ostentation. They were also men who sang in public the praises of free enterprise and individual initiative, yet in business they acted in associational ways; they worked hard to undo what they perceived as ruinous competition and helped usher in an age of organized capitalism. They similarly gave to charity, but normally showed no mercy to their workers. Their varied business practices and contradictory lives do not allow for easy labelling—as either great innovators or robber barons, for example. On balance, however, a negative judgment sticks, for these men, with perhaps the exception of Andrew Carnegie, never once stopped to consider how their rapacious activities in the marketplace, their traducements in the political realm, and their building

of large-scale instititutions threatened treasured democratic republican
visions and values of the American people. Ironically for them, too, historians in recent times, despite the obvious fascination these men still
hold and the good copy they provide, have even downplayed their very
place and contribution in the rise of big business.

The Managers

Caution is in order when attributing great agency to the famous business figures of the late nineteenth century for one simple reason: many of
the large-scale enterprises formed in the period failed to survive. Standard Oil and General Electric secured a permanent footing, but entering
the dustbin of history were scores of other such ventures: National Cord-
age, United Copper Mining, Great Western Cereal, Consolidated Rubber
Tire, American Cement, American Bicycle, U.S. Leather, U.S. Shipbuilding, and United Button, to mention a few. A number of these would-be
giants were not meant to succeed; they were conceived by speculators
who had no intention of establishing and managing ongoing concerns,
but rather, aimed at immediate killings in the stock market. Technology
played a role, too. Big businesses tended to prevail in industries where
standardized goods were produced, where machines could easily replace
hand labor, and where economies of scale and throughprocessing were
achievable—for example, in petroleum, plant oil, chemical, sugar, and
alcohol distilling and refining; iron, steel, copper, and aluminum manufacture; and grain and tobacco processing. Large-scale companies typically did not appear or succeed in apparels, textiles, shoes, lumber,
furniture, leather, machine tools, and printing. Changing, small-batch,
custom orders dominated in these trades and were not well handled in
the setting of the large enterprise.

Technology alone, however, did not determine the survival of big
businesses. The potential may have existed to produce and market a good
at low cost on a large-scale basis, but full realization of savings and
profits could only be reached with innovative and able management. The
short histories of a number of the great enterprises formed in the late
nineteenth century can be attributed to the speculative intentions of their
initiators; the failure to forge effective administrative structures and practices doomed many other ventures. Praise and attention are due to the
appointed managers who transformed possibility into performance, who
made the companies founded and set in motion by the notable businessmen of the age work. They were principal agents in the rise of big business as well.

The life stories, even the names, of the men who helped sustain the new giants are lost to the historical record in many instances. Scholars know more about what they did than about who they were. The appointed managers made their mark first in charting basic sales strategies for the new firms and then in developing appropriate organizational schemes, production systems, accounting procedures, company rules and regulations, personnel practices, and feedback and forecasting methods; later, they would staff and administer the concerns they so carefully molded.

The first generation of corporate administrators operated in unchartered waters, relying on intuition and trial and error, but they did have an important precedent to guide their efforts: the American railroads. The railroads may not have been the prime movers in early American industrialization, but they did play an important role in the introduction of bureaucratic principles and techniques of management. By the 1850s and 1860s, railroad managers had already made great breakthroughs in the administration of large-scale enterprises, and they bequeathed valuable examples and lessons to the appointed directors of the new industrial behemoths created forty and fifty years later.

When the first American railroads commenced operations in the 1830s, the prominent stockholders who comprised the boards of directors divided managerial tasks among themselves. As service expanded, it became clear that the founders of the railroads, who most often were merchants by trade, had neither the time nor the expertise to see to the daily running of the trains. The boards then moved to hire others to manage their properties. Appointed administrators at first assumed direction of all activities on the roads, but this soon also proved infeasible with the extension of operations over great distances and the growth of employment. More detailed supervision was required, and additional managers were engaged to separately oversee the maintenance of the roadbed and tracks, repair of engines and rolling stock, the dispatch of trains, and the workings of the stations and yards.

The first corps of railway managers were recruited from the construction, machine, and civil engineering trades. Here, the railroads forged a link between the engineering profession and management that persists in business to this very day. The railroads would also develop new internal promotion systems to fill administrative positions. The great corporations constituted in the late nineteenth century followed the early railroads in recruiting professional engineers for managerial posts and in promoting from the inside, but the latter-day giants also began to secure administrators from new graduate schools of business administration founded in the same age.

Expansion of rail service required greater numbers of supervisors and also greater attention to procedure. In the 1840s, appointed managers on the Baltimore & Ohio Railroad led the way in the writing of lengthy manuals to guide all aspects of operations. The B&O protocols detailed the tasks and responsibilities of every employee, stipulated clear lines of authority, and established a system of reporting where each level of management accounted for its activities. Notable in the railroad's original handbooks were provisions for new financial officers of the company; these appointed managers would handle external funding and check the internal flow of revenues. Extensive rules were included for the collecting of fares, receiving and receipting of freight, and transmitting of cash. By the late 1840s, salaried bureaucrats had basically assumed complete control of operations on the B&O.

The handbooks of the early railroads indicate that their managers had opted for a functional approach to administration. Department heads oversaw different activities on the lines. In the 1850s, renewed expansion and dispersion of service forced managers on such major carriers as the B&O, the Erie, and the Pennsylvania to develop new organizational principles. Daniel McCallum, general superintendent of the Erie, initiated the first changes. The railroad then had more than four thousand employees and stretched through the state of New York, and there was a pressing need for new managerial strategies. (The Erie continues to capture the attention of historians for the financial escapades of Jay Gould, and only with the recent shift in focus to the role of managers in the rise of big business has its place in administrational history been recognized.)

To establish greater controls on the sprawling enterprise, McCallum decided to divide the line into four geographical regions. Over each, he appointed divisional superintendents who received complete responsibility for the day-to-day movement of trains and the upkeep of machinery and track within their domains. They in turn appointed regional department heads to oversee different tasks. McCallum also created a central staff of departmental officials to coordinate activities across the geographical divisions of the road. To enforce discipline, he gave the power of hiring and firing of subordinates to immediate supervisors and instituted a system of detailed hourly, daily, and monthly reports that became standard in the industry. The notion of administrative hierarchy was so important to McCallum that he drew up and had lithographed for public distribution a precise organizational chart of the Erie's operations, probably the first of its kind for an American enterprise.

McCallum, however, never solved a critical problem with his system—an unclear relationship between central department heads and divisional officers. Administrators on the Pennsylvania Railroad in the later 1850s then improved upon McCallum's scheme. In a managerial plan devised in

1858, the Pennsylvania, which would soon emerge as the largest corporation in the world in terms of both income and employment, was divided into geographical divisions. Regional superintendents and their department heads were delegated absolute and definitive charge of all operations in their provinces. Department heads at central headquarters, on the other hand, received authority to develop plans and overall strategies, set standards and procedures, make inspections, and advise divisional officials. They were afforded no direct role in the daily supervision of operations and employees.

The Pennsylvania plan thus established a clear differentiation between line and staff officers, between day-to-day decision makers and long-term organizational planners. Ambiguities in the Erie system were eliminated. In decentralizing authority and the supervision of activities, the Pennsylvania design also served to create small, relatively autonomous units within the one mammoth enterprise. With minor variations, other railroads adopted the line and staff–differentiated administrative approach of the Pennsylvania. Further managerial breakthroughs occurred on American railroads in later decades. Railway managers perfected new methods of ascertaining the costs of the various tasks involved in rail transport, so that they now had better information for the setting of rates; when the carriers began to cooperate and pool their resources and share freight and passenger traffic, ending their competitive warfare, rail officials also developed effective cross-accounting and reimbursement procedures. Still, by the time of the Civil War, appointed managers of American railroads had already concluded basic experimentation in the administrative design of large-scale enterprises.

The railroads offered valuable lessons for the governance of big business: the successful firm would separate ownership and administration; professionalize management; create clear hierarchies, rules, and procedures; attend to cost accounting; divide operations either by geography or by product line; delegate broad powers to division heads and their departmental lieutenants; allow the divisions great autonomy; create a central staff for overall coordination and planning; keep central department officers focused on future developments and not on day-to-day problems; and cooperate with the seeming competition. The railroads also offered evidence of the gains to be achieved through cooperation with trade unions and government agencies.

The managers placed in charge of the new large-scale enterprises of the late nineteenth century had the ready example of the railroads to follow. Still, the development of the administrative schemes that helped sustain the companies established by the great business figures of the age was neither easy nor immediate. Improvisation was required. More important, conflict marked the process and forced constant alterations. There were internal tensions. The founders of the new firms and their families only reluctantly relinquished family control over management; acting conservatively, they resisted innovations suggested by the hired professionals. Appointed administrators also fought turf battles among themselves, and labor conflict in some instances forced new standards and procedures on unwilling bureaucrats.

External pressures also affected developments. With the possible exception of the aluminum industry, where the Alcoa Company emerged as the single operator in the trade, none of the giants created at the turn of the twentieth century succeeded in achieving a monopoly position. The market and the resources of the country were too large for competition to be entirely eliminated. John D. Rockefeller would soon face upstarts in the oil supply and refining business after discoveries of vast deposits of oil in Texas and the Southwest; foreign companies also challenged his realm. Typically, as in oil and steel, five or six firms came to dominate in specific industries where there were large-scale firms. Persisting competition forced administrators to develop new sales strategies and revise their management systems. The new large enterprises especially had to learn how to deal with the problems of overproduction and excess capacity. Slowly, they moved to diversify and expand their product lines.

The classic example here is provided by the Du Pont Company. Until World War I, this major chemical company specialized in the manufacture of gunpowder and explosives. The 1920s found the firm with a depleted market for its product, and managers at Du Pont embarked on an ambitious plan to use the chemical resources and expertise of the company to produce a variety of consumer goods from paints to synthetic textile fibers. Diversification required restructuring the firm along the divisional, line and staff–differentiated system of management developed by the railroads. Diversification became an important answer for many companies in dealing with overcompetitive markets, and this hastened a general move to the division and decentralization of management. Finally, administrators of the new firms not only faced external pressures from other businesses, but also from popular opposition to the corporations and from new government regulatory agencies; methods had to be revised in reaction here as well.

Change in business practices, then, was ongoing. Extraordinary business figures established mammoth enterprises in the late nineteenth century, yet the success or survival of these organizations was not guaranteed. The new giants required apt administrational systems. The appointed managers of these firms had the flowcharts and accounting methods of the railroads as a reference, but through circumstance and internal and external pressures, the fixing of clear and effective sales programs and managerial schemes took time and alteration. This by no means depreciates the role of the managers in the rise of big business. In realizing the potential of the large-scale enterprises, the managers figured as crucially as the more headline-capturing founders.

The Role of Events and Circumstances

The feats of the great business figures—be they the founders or the managers—do not provide sufficient explanation for the rise of big business in the United States in the late nineteenth century. Attributing developments to Great Men always raises questions. Why in this particular time and place did such an extraordinary cohort of individuals suddenly appear to change the course of history? Greatness, one assumes, is distributed randomly across time and space. Were there not special circumstances that allowed these eager and able men to achieve as they did? Had conditions been otherwise, would their accomplishments have been as daunting? The emergence of big business can in fact be ascribed to any number of contingencies.

The Political and Legal Framework

The American political and legal systems encouraged large-scale enterprise, though in odd ways. The corporation had been legally accepted at the dawn of the republic, and the principle of limited liability was legally recognized by the middle of the nineteenth century. The decentralized political structure of the country allowed for eased access to incorporation rights. Ironically, when antimonopolist sentiment in the 1830s led to the end of the special legislative grant of corporate privileges, the states enacted general incorporation laws and simple administrative procedures for securing charters that further promoted the corporate form of business.

In the late nineteenth century, other legal and political developments hastened large enterprise. Under the pressure of intense competition, businessmen starting in the 1870s looked to curtail their rivalries through cooperative agreements. Firms entered into secret pacts to carve up market territories and set floors on prices. This often occurred within the context of newly established trade associations that emerged in industries and communities throughout the industrial North.

No means existed to enforce the accords reached by rival businessmen, and invariably they collapsed as individual firm owners reneged on their commitments to joint action and decided to protect and boost their own interests. A renewed search then often ensued to find a different way of forestalling injurious competition, and the next resort was to the holding company. John D. Rockefeller provided the first evidence of its efficacy with the formation of Standard Oil in the mid-1870s. Rockefeller convinced many of his competitors to come under the wings of this entity and abide by the decisions of its board of directors in return for cash or Standard Oil stock. Problems still emerged with this arrangement. Technically, the parties to the agreements remained independent and could withdraw and induce a new round of price-cutting. More important, the legal standing of the holding companies was greatly in question. The agreements reached among businessmen to diminish competition, whether informally or through trade associations and holding companies, represented conspiratorial acts to restrain trade and do harm to others and were thus illegal under common law traditions.

The nation's divided system of government, however, left matters in limbo. Since incorporation rights remained a prerogative of the states, federal mandates did not exist at the time to effectively prosecute the collusive actions of businessmen. Political conflict at the state level delayed clear legislative action, and as a result the issue rested in the hands of local judges, who issued inconsistent rulings. Although few judicial decisions were issued against the collective decisions of firm owners, the threat still existed. For this reason, Rockefeller experimented with the idea of the business trust in 1881. Discipline of the constituent firms appeared more secure under this arrangement, since the company owners literally handed their deeds and stocks to the trustees; as an estate, Standard Oil now seemed immune to charges of conspiratorial behavior.

The founders of other large-scale enterprises followed Rockefeller's lead in establishing trusts, but the legal issue was still not settled. Definitely questionable was whether trusts officially created in one state could hold the stocks of companies chartered in another. Once again, the nation's decentralized political system would play a role in developments. Just as state control of incorporation rights encouraged corporate formation in the antebellum period, localism also promoted conglomerate corporate efforts, holding companies, and trusts in the late nineteenth century. These more complicated enterprises required special legislation—they did not fall under general incorporation laws—and a Jay Gould could hold sway among legislators in New Jersey; a Thomas Scott, president of the Pennsylvania Railroad, easily in Pennsylvania; and John D. Rockefeller in Ohio, to see to the chartering of their consolidations. Equally important, with decentralization, competition prevailed among states to entice business; as a result, there was pressure for the writing of permissive legislation. Here, New Jersey assemblymen proved most willing. In

1889 they modified the state's general incorporation laws to basically legalize the trust. The new statute permitted manufacturing companies to purchase and hold stock in other companies within and outside the state. The New Jersey law was a definite inducement for business confederations.

The full history was still not written. Just one year after New Jersey legalized the trust, congressmen in Washington, bowing to mounting popular opposition to the seeming conspiratorial and monopolistic actions of the new business kings, unanimously passed the Sherman Antitrust Act. The law explicitly forbade all business "combinations in the form of a trust or otherwise in restraint of trade." The federal government had stepped into the legal breach on the basis of federal jurisdiction in interstate commerce. Ironically, the Sherman Act would not halt business conglomeration but merely bring the process to a new stage; specifically, the law spurred businessmen to move from associative arrangements to pure and simple mergers.

The actual legal consequences of the Sherman Antitrust Act remained hazy. New Jersey legislators changed the wording of their new incorporation law to allow once more for holding companies. The Supreme Court, in a first test of the Sherman Act, found New Jersey's revised legislation in compliance with the federal mandate, but in subsequent decisions in the late 1890s the Court ruled that any confederation of business firms formed to fix prices or allocate markets was in violation of the antitrust law. In the next twenty years, the court would wend its way toward a fixed but middle position: business combinations were not illegal per se; the issue would be the means—"reasonable" or illicit—by which market controls were achieved.

The concrete result of the Sherman Antitrust Act and attendant Supreme Court decisions was uncertainty. Alliances of any kind among firms risked prosecution, and businessmen realized and were advised by their lawyers that in trying to curb competition, their only secure approach was the thorough blending of constituent companies into singular concerns. Merger loomed as the answer, and not coincidentally, a merger frenzy began to take hold in the mid-1890s. At its peak, between 1894 and 1904, 131 mergers were recorded, involving the complete disappearance and subsuming of more than 1,800 companies into new, consolidated enterprises. The threatened if not absolute outlawing of cartel arrangements thus bred mergers in the United States. In contrast, cartels remained legally acceptable in European countries such as England and Germany. There, informal and formal alliances among firms to stabilize markets remained the norm, not mergers.

Thus, the American political and legal systems encouraged early and continuing use of the corporate form of enterprise and later experiments with holding companies and trusts: decentralized government increased access to important privileges. Divided government also left great leeway for associative business activity. Eventual prohibitions against cartels and ongoing legal uncertainties, then, promoted mergers. Ironically, anti-monopoly politics had the unanticipated consequence of furthering business concentration throughout the nineteenth century. General incorporation laws of the antebellum period generated more corporations; later protest against business conspiracies and collusion begot mergers. Antimonopolism encased big business in one more instance. In the early twentieth century, the Supreme Court would declare such companies as American Tobacco and Standard Oil in violation of the Sherman Antitrust Act and order their dismemberment. Typically, the firms were broken into separate new concerns, but these new large-scale entities themselves came to dominate their industries. In effect, antimonopolism led to oligopoly, or the control of trades by a few firms, not to truly competitive marketplaces. The political and legal order in America thus contributed to and shaped the rise of large-scale enterprise over the course of the nineteenth century and particularly in the last decades of the century.

The Depression of 1893

The formation of big businesses in the latter part of the nineteenth century transpired in stages. Throughout the period, enterprise builders such as Gustavus Swift steadily enlarged their firms vertically through the assumption of new activities; but for the large enterprises formed horizontally through the absorption of similar firms, there was a definite steplike process. Continually propelled by the need and desire to limit competition, they progressed from the informal pacts, trade association agreements, holding companies, and trusts of the late 1870s and the 1880s to the merger movement of the late 1890s. The rise of big business is often associated with the mergers of the last decade of the century, but mergers were a final phase in but one path toward large-scale enterprise. The period of the mergers was so frenzied and involved so many firms and such large transactions that it is rightfully identified as a key moment in this history.

Legal decisions provide the backdrop to the merger movement of the late 1890s. Prohibitions against cartels left merger as the only option for curbing competition. The major consolidations of the decade cannot be understood without an appreciation of the legal underpinnings. Other precipitating events, however, have to be added to the story. One of these is the so-called depression of 1893, an immediate and severe economic contraction that began in February of 1893 and did not lift for more than four years. There were various causes for this great depression—economic difficulties in Europe, a peaking in railroad construction, a troughing in a normal business cycle—but the most proximate was the bankruptcy of several important firms and investment houses. The depression of 1893 had great social consequences, with massive unemployment and personal suffering, and eventually great political ramifications; it also proved disastrous for those manufacturing firms that had recently built up impressive capacities and had taken on huge fixed debts in the process.

The economic downturn triggered a merciless price war as businesses tried to raise revenues just to pay off their creditors. Companies pulled out of existing cartel arrangements, many folding and looking for buyers. The depression itself bolstered the notion of mergers—as if the legal messages were not sufficient. No cooperative arrangements could withstand another round of price-cutting. Individuals with resources could also now purchase facilities on the cheap and incorporate them into large, single firms. J. P. Morgan was one of these; during the years of the depression he was particularly busy forming new amalgamations, particularly in railroads. The depression thus played its part in ushering in the merger stage of large-enterprise formation.

The Changing Role of Finance Capital

The example of J. P. Morgan points to a third underlying element in the merger movement of the 1890s, one beyond law and immediate economic circumstances. Investment bankers like Morgan had assumed no role in the building of large industrial firms during the last decades of the nineteenth century. Neither a Rockefeller nor a Swift sought the services of Morgan and other securities dealers; they brought in partners and borrowed money directly, usually from local sources, when raising funds for expansion. Wall Street financiers, in the meantime, remained content to specialize in the circulation of railroad securities, government bonds, currency, and precious metals.

In the early 1890s, circumstances changed for them. With railroad construction reaching an apogee, the sale of railroad stocks and bonds, their main business, no longer offered opportunities for profit. Investment bankers then began to look at the market for industrial securities. Their customers welcomed new investment outlets; the financiers could convince the industrialists of the efficacy of issuing securities and enhancing their total assets. The investment bankers, in fact, egged on the merger movement. They brought the parties together, handing out huge blocks of new stock to the sellers of troubled firms, to those taking charge of the new consolidated enterprises, and to themselves, and marketing the rest to the public (of course, for additional commissions). Capital was raised for the operation of the new merged businesses, and with the proper manipulation of publicity, everyone holding the securities of these firms could be greatly enriched. The merger movement occurred partly because finance capitalists had much to gain, and here, changes in the securities industry in the early 1890s with the culmination of railroad construction in the nation played an important role.

Labor Unrest

One other circumstance to be considered in explaining the great business mergers of the 1890s relates not just to the merger movement but to the general rise of big business in the late nineteenth century. The same age that saw the emergence of large-scale enterprise was also witness to extreme conflict between labor and capital. The industrial strife of the period is often depicted as a response to corporate growth. But labor unrest also contributed to business consolidation and was agent and product of change.

Throughout the nineteenth century, skilled workers in industrial concerns maintained great control over the pace and quality of production. Fierce competition within trades required close attention to costs, and the high wages of the skilled men emerged as one critical matter. For large firms, profits rested on volume sales, and utter efficiency on the shop floor was an imperative; here, the skilled workers' technical knowledge and command over the pace of operations loomed as equally problematic for company owners. Cooperation with other businessmen in stabilizing prices appeared to be a means of dealing with the high costs of labor; this is one link between capital-labor conflict and the move toward confederated enterprise. The elimination of skilled workers and their replacement with machinery and cheap hands emerged as the solution for managers for gaining effective control in the workplace. This move required great capital resources, often affordable only through the consolidations engineered by financiers. A line between labor tensions and mergers can thus be drawn.

Labor unrest, in fact, served as a backdrop to the rise of big business. Throughout the 1880s and early 1890s, key battles were fought in the steel and farm machinery industries, among others. These confrontations involved managerial attempts to defeat the unions of skilled workers. Whether lost or won by management, they were invariably followed by efforts to raise capital, business consolidations, increased division of labor, mechanization and automation, greater supervision, release of skilled workers, and the hiring of unskilled immigrants. The connection be-

tween labor conflict and big business was described well in this respect
by Charles Flint, founder of the U.S. Rubber Company. Writing in 1901,
he noted that "the American workingman produces more [today], and he
produces more because he has been supplied with the most perfect system of labor-saving machinery on earth. To supply this machinery, large
capital is necessary. The individual manufacturer, standing alone, is not
in a position to perfect his machinery in the same measure as the consolidated enterprise." Again, context is important. The emergence of large-
scale enterprise is not understood without due appreciation of the simultaneous crisis of relations on the shop floor.

Forces

Business figures—founders and managers—did not operate in a separate, insulated realm. Events and conditions around them forced their
hands and contributed to the nation's particular history of big business
formation. There is a way of treating the rise of large-scale enterprise,
however, that downplays the role of seemingly key individuals and circumstances. The new, encompassing firms of the late nineteenth century
can also be seen as the product of long-range, impersonal demographic,
market, and technological forces.

This view can begin with the vast expansion of the American population that occurred in the nineteenth century primarily through continuing, massive immigration. (Fertility declines marked the period.) The
newcomers swelled urban areas, where they remained reliant on the market for livelihoods and goods and services; those who settled on the land
entered, with few exceptions, into commercialized agriculture. (The rich,
arable land base of the country and the extraordinary productivity of
American farmers allowed also for increased urbanization and industrial endeavor.) Rapidly expanding markets acted as an inducement for
mass production in the context of the large-scale enterprise.

Add the availability of capital for the building of huge manufacturing facilities to the equation. With a less than well-established banking system, the
nation did face capital shortages. Industry expanded through the taking in of
partners and the turning of all savings and profits into new capacities. Yet,
the investment bankers who orchestrated railroad mergers and later industrial consolidations had no difficulty in marketing corporate bonds and
stocks (at home as well as abroad) for the raising of funds. Surplus monies
existed for the financing of big enterprise, and the source was usually families who had accumulated great wealth in commerce. Merchant capital thus
promoted and bankrolled industrialization in early and later stages.

A growing market and available investment funds are two basic elements in the rise of big business. The building of an infrastructure of transportation and communications services is an essential third. Canals and railroads (and the telegraph) extended the geographical range of the market and allowed market activity to expand. There is an important and specific link, however, between these developments and the rise of big business. With a national transportation and communications system in place, no firm in the United States by the late nineteenth century operated in isolation, with known and reliable customers. Companies now faced inescapable competition, and this forced them to act in new ways. Expansion and extension of operations represented essential responses. There is thus a technological component in the emergence of large-scale enterprise—new transportation and communication technologies served as the base of change—though this history would not have unfolded without a growing population of customers and adequate sources of investment.

Thus, demographic, market, and technological forces generated large-scale enterprise. Hidden in this generalization, though, is a more detailed story (but one not requiring mention of famous businessmen or specific events). Competition, induced by the creation of a national marketplace, caused firms to grow. They expanded in two different ways, horizontally and vertically. Horizontal expansion, as we have already seen, occurred in the late nineteenth century. Faced with competiton, companies typically in the 1870s sought to reach agreements among themselves to allocate markets and set floors on prices within industries. Pacts proved difficult to sustain, and notable in the period is a progression from informal accords to trade association compacts and then to experiments with holding companies and trusts. Ultimately, the only solution resided in absolute merger (although politics and law played a role here). Competition, unleashed with the creation of a nationally based market by transportation and communications improvements, thus drove horizontal expansion. Classic examples can be found in such industries as oil and steel.

Vertical growth provides a different narrative. Faced with competition, some firms decided to control their supplies of raw materials. Trying to produce on a mass basis, they chose not to remain in the grips of outside suppliers, who could determine prices and the flow of resources to the firms. Manufacturers thus integrated backwards—creating new departments to handle the accessing of materials, growing in the process— although this was not always wise or necessary. If competition reigned among suppliers, producers could scout for the best deals rather than take on new administrative costs. But where manufacturers remained at the mercy of single provisioners, assuming management of supplies was an imperative.

Competition also drove firms to delegate greater time and effort to merchandising. Before the 1880s, manufacturers typically relied on commission salesmen to market their goods. With no customers now guaranteed, given extended transportation and communications facilities, firms had to take full control of distribution. This required the establishment of other new offices to manage advertising, contacts with wholesalers and retailers, ordering, warehousing, and distribution. Firms thus grew additionally through a process of forward integration. But there was a technological aspect to the growing emphasis on marketing as well. Companies often had to invest great resources in sales because they were producing novel products. The public at large had to be informed about their use and convinced to change habits and buy—whether the product was canned soup, dressed beef, or electricity. In Gustavus Swift's meatpacking concern, the size of the sales force began to rival the numbers of workers in production.

Few firms grew either just horizontally or vertically. As with Andrew Carnegie's steelworks or the American Tobacco Company, firms expanded both through extension of activities—directly accessing materials and distributing products—and through various kinds of horizontal combinations. Still, distinctions are in order, for firms can be designated as expanding either more vertically or horizontally. Whatever the actual process, the rise of large-scale enterprise can be understood as the result of basic demographic, market, and technological forces.

The rise of big business, then, is a complicated story with various ways to understand developments. A comprehensive perspective is possible, however, with weights assigned to different elements. An expanding U.S. population and an extensive transportation and communication infrastructure propelled competition and forced firms to act in new ways to horizontally and vertically integrate. New products produced by new technologies demanded greater attention to the marketing of goods, which also fostered expansion of operations. New labor-saving machinery was affordable and efficiently operated in the context of well-endowed, large firms. The availability of investment capital facilitated conglomeration. Most of the story appears to lie in such impersonal considerations. Moreover, the particular paths taken and the timing of change can be attributed to special events, such as the outlawing of cartel arrangements and the impact of the depression of 1893. This, too, leaves people out of the picture, or at least on the sidelines.

What, then, of the great business figures of the period? Where do they ultimately fit? Definite weight has to be assigned to the managers. They made the new enterprises work (to be sure, after a great deal of trial and error) with innovative organizational and accounting techniques. The great enterprise builders, on the other hand, remain somewhat of a problem in this puzzle. Would the awesome industrial works that appeared in the United States in the late nineteenth century have been planted without the likes of a John D. Rockefeller or a J. P. Morgan? Probably. The impetus was there for big business. But these men cannot just be dismissed. They acted so relentlessly and voraciously that it is impossible to see developments unfolding as fast and on such a scale without their presence. Assign them a part—but not a leading one—in the rise of large-scale enterprise. Give them a greater role, perhaps, in the explosive response of the American people to the fixing of the powerful corporation in the American landscape. As symbolic figures, the so-called robber barons may ultimately have had a greater impact on the nation's politics than on its business.

Chapter Seven

Explosions

Social Unrest in the Late Nineteenth Century and the Remaking of America

MONDAY, JULY 16, 1877. Martinsburg, West Virginia. A date and place emblazoned neither in history books nor in the historical consciousness of the American people. Yet, on that day and in that location, the people of the United States stepped precipitously into the future.

On July 16, 1877, railway workers in Martinsburg, employees of the Baltimore & Ohio Railroad, refused to handle rail traffic or let trains pass through the town. They were protesting the implementation of a 10 percent cut in wages that had been announced simultaneously a few weeks earlier by railway executives of the major rail lines in the country. The concerted nature of the announcement would be an important element in the story to unfold. In response to the job action, the president of the B&O persuaded the governor of West Virginia to send regiments of the state militia to Martinsburg to see to the safe movement of trains. The easy access of corporate leaders to the levers of government power would be an additional ingredient in the saga. To the dismay of B&O officials, however, the troops who arrived on the sixteenth initially fraternized with townspeople. Later that day a melee did erupt, and in the ensuing fracas a striking railwayman guarding a track switch was shot and killed. That proved to be the spark that ignited a nationwide conflagration.

Word soon spread along the tracks of the B&O, and work and traffic on the entire line ground to a halt. On Wednesday evening, angry railwaymen and their supporters gathered in Baltimore to protest directly to B&O officials. Protest turned to riot; and by night's end, 10 people were dead, 16 injured, and 250 arrested through confrontations between demonstrators and city police. Shocked by the insurrection, the governor of Maryland prevailed on the president of the United States, Rutherford B. Hayes, to send federal troops to Baltimore to quell the disturbance—the first time in American history that federal forces were employed to suppress labor unrest.

The fire then spread to other communities in the country. Following the example of B&O workers in refusing to accept wage cuts, railwaymen from other lines walked off their jobs and were joined by fellow townspeople in demonstrations. Protests soon emerged in such places as Hornelsville and Buffalo, New York, and Reading, Harrisburg, and Altoona, Pennsylvania. The greatest explosion, however, was to occur in Pittsburgh. On Thursday, July 19, railwaymen from the Pennsylvania Railroad stopped rail traffic in the city. On Friday, state militia from the area were called in at the request and insistence of Thomas Scott, influential president of the road, with the aim of restoring train service. The local guardsmen, however, refused take up their posts. Troops then had to be sent in from other parts of the state—a move that residents of Pittsburgh perceived as an invasion—and this set the stage for a brutal confrontation on Saturday. Pennsylvania Railroad executives were determined to renew freight traffic, and they arranged for troops to be stationed on the trains. As the first guarded train moved through the city, crowds gathered to block its progress. Troops then fired into the crowd of demonstrators, killing an estimated twenty people and wounding more than seventy more. Word of the massacre quickly spread, and the people of Pittsburgh took to the streets, attacking militiamen, looting stores, and setting fires to the property of the Pennsylvania Railroad. By late evening of Saturday, July 21, 1877, a red glow lit up the city; and daybreak revealed the stations and shops of the Pennsylvania reduced to embers.

Sunday did not prove to be a day of rest. Crowds continued to roam the streets, and a semblance of order would not be restored in the city until Tuesday. Over the weekend, more than two score lives had been lost, 104 locomotives and 2,153 railcars had been destroyed, and few buildings of the Pennsylvania Railroad remained standing.

Chicago was next. Three days of serious disturbances began in that city on Monday, the twenty-third, and confrontations between protestors and police would lead to eighteen dead. The railroad strikes in Chicago evolved into a general strike as workers across trades in the city walked off their jobs in sympathy. Order was restored only with the arrival of Illinois guardsmen from other parts of the state and a contingent of federal troops just fresh from fighting Native Americans on the Plains. After Chicago, the contagion spread further west to St. Louis, Kansas City, Galveston, and even San Francisco. Only two weeks after its onset did the fever run its course. Trains first rolled again on a normal basis through Martinsburg on July 27, through Pittsburgh on July 29, and through Chicago on August 3.

The events of July 1877 shocked the nation, and the toll was enormous. The nation's commerce had been effectively stilled; railroad companies lost more than $30 million in lost property and business; railwaymen went without pay during the strikes, and many returned to work only to be discharged for their protests and blackballed from further employment in the trade; thousands had also been jailed, hundreds wounded, and at least fifty killed.

From the last two decades of the nineteenth century through World War II, ongoing and vexing conflict between capital and labor marked the American experience. The great railroad strikes of July 1877 represent the formal and abrupt beginning to this history. In the immediate years following the unrest of that month and until the turn of the new century, industrial strife was particularly intense and violent. The severity of the unrest can only be understood as a twofold response: while economic insecurity, not hardship per se, definitely spurred revolt, Americans from different walks of life also took to the streets to support striking workers in the period to challenge the growing, encroaching political and economic power of concentrated capital and the threat the corporation posed to cherished democratic republican values and practices.

During the last two decades of the nineteenth century, the country was also rocked by other kinds of explosions. Most notably, the period witnessed evolving and escalating protest by American farmers. Various social critics and reformers also came to the fore to question contemporary developments, galvanize public opinion, and suggest changes in economic and political practices. The economic instabilities of the times and the sway of the new corporations figured significantly in the complaints of farmers and intellectuals as well.

Unrest during the last decades of the nineteenth century would lead to a restructuring of American institutions and the creation of a new American political economic order. This remaking, however, occurred slowly over a fifty-year period, in stages and unsystematically, and would involve no single set of actors. Various groups would emerge and, often with cross-purposes and differing motives, contribute to the same building of a new United States. What is of interest is the *convergence* of efforts. What was sought in common was greater security and a more administered economy and polity.

Labor Unrest

The great railroad strikes of July 1877 were the first salvos. In the next two decades, the nation would be shocked by a succession of other monumental battles between capital and labor. In 1885, for example, railwaymen working on lines controlled by the infamous Jay Gould walked off their jobs in protest of announced wage cuts, shutting down most rail traffic west of the Mississippi. Under financial pressure, Gould capitulated to the strikers, a David-and-Goliath victory that would be celebrated in working-class communities throughout the country. The strike had been fought under the banner of the Knights of Labor, and Gould's defeat would boost the fortunes and visibility of that relatively new trade union federation. Gould would have his revenge, however; a second paralyzing and more violent strike launched a year later by railwaymen on his lines would conclude this time with labor's capitulation.

The dust had hardly settled from the second so-called Gould strike when the nation was rent asunder again by worker protest. Throughout the 1870s and early 1880s, various trade union leaders had made the eight-hour workday a central demand of labor (twelve hours of work a day was the norm at the time). A nationwide general strike on behalf of shorter hours of work was announced for May 1, 1886. An estimated 200,000 workers responded to the call, closing down shops and mills in small towns and major cities, and attending rallies where speakers proclaimed that shorter hours would allow workers time for well earned rest, but also time to pursue their own interests, education, and the fulfilling of their civic duties as citizens of the republic. A few skirmishes erupted between police and demonstrators, but a major incident was to occur in Chicago. On May 3, eight-hour supporters joined strikers at the McCormick Reaper works for a rally. Chicago police, fixed on breaking up the demonstration, waded into the crowd, shooting and killing four strikers in the process. A protest meeting was then called for that night at Haymarket Square. Between 2,000 and 3,000 people attended what at first was a peaceful gathering, but as they later dispersed, a bomb exploded in the midst of a contingent of policemen. Eight officers were killed, and as other police responded with gunfire, blood flowed in the streets of Chicago—with eight in the crowd killed and more than fifty wounded.

The Haymarket bombing reverberated throughout the nation. Police in Chicago clamped down on all protest activity, aided by state militia and a corps of deputized property holders. Furthermore, the police detained hundreds of people suspected of radical sentiments or membership in radical organizations and finally charged eight proclaimed anarchists with the bombing. At a subsequent sensational trial, the prosecuting attorney successfully convinced a jury (which included not a single worker) that while the actual guilty party could not be determined and six of the eight defendants could not even be placed at the event, all eight were guilty of conspiracy in the placing of the bomb. As a result of the convictions, seven of the accused were sentenced to death by hanging. The case would remain in the public mind for years. A protest movement emerged to demand a retrial. Despite such opposition, four of the seven convicted were hanged, one died in prison, and two others had their sentences commuted to life imprisonment. The Haymarket tragedy further heightened tensions in the nation and made labor unrest a central issue of the day.

The Haymarket tragedy also had the deleterious effect for the trade union movement of causing labor organizing to be identified in the public mind with radicalism and incendiary activity. It did not force a retreat, however. Railwaymen took up the gauntlet again. In 1888, locomotive engineers and firemen on the Chicago, Burlington & Quincy launched a strike that spread to other lines and threatened anew the nation's commerce. In 1894, a dramatic strike would be won by workers employed on the Great Northern Railroad, a line controlled by the financier James Hill; this battle would thrust a young, charismatic labor leader, Eugene Victor Debs, into the national limelight.

Steelworkers also joined the fray. In 1892, employees of Andrew Carnegie's steelworks in Homestead, Pennsylvania, participated in one of the most legendary strikes of the period. In an effort to seize complete control over production, Henry Frick, Carnegie's associate and general manager, determined to expunge the strong craft unions from the plants. He declared war on the Amalgamated Association of Iron, Steel, and Tin Workers, the strongest and best-organized union in the country. In late June of 1892, at the termination of a contract with the Amalgamated, he announced his refusal to deal further with the union. Frick then built fortifications around the Homestead works, instructing guards not to allow Amalgamated men into the plant; the guards were also there to protect newly hired nonunion men. Frick needed greater protection, however, and this set the stage for a pitched battle.

On July 6, three hundred private police from the Pinkerton National Detective Agency arrived by water near the Homestead plant on covered barges. Workers locked out of employment immediately attacked the invaders, pelting them with stones and bricks and firing guns. For hours, the defiant steel men and the Pinkertons exchanged shots. An armistice was eventually arranged and the private police force allowed to land, but not before nine steelworkers and seven Pinkertons lay dead. The state militia soon arrived to restore order, but also to allow Frick to hire more nonunion men.

The Homestead strike continued to capture attention throughout the summer. Striking workers attempted to extend the protest to other facilities; there was a dramatic attempt on the life of Frick; union men were soon rounded up for trials, accused of conspiracy, rioting, and murder; and the Amalgamated became more immobilized. In the fall of 1892, Carnegie and Frick resumed full production with a lessened union presence as they continued their effort to take controls on production away from the skilled men. Bitterness would prevail in the community of Homestead for decades.

The epic Pullman strike and boycott of July 1894 capped the great labor disturbances of the last decades of the nineteenth century. George Pullman had become famous in the 1870s for the manufacture of his sumptuous railway sleeping and dining cars. He also had attracted attention in the 1880s for building a seemingly model community just south of Chicago for families of the men who labored in his shops fabricating Pullman cars. Harmony in this well-landscaped and complete company town, though, was just an appearance.

In June of 1894, Pullman announced a reduction in wages; a severe economic depression that had begun a year earlier had demanded cost cutting. Employees of the company walked off their jobs in protest. Pullman had refused to lower rents in the already high-cost lodgings that he provided his workers, so the wage cuts represented a serious hardship. Pullman reacted to the strike by closing down his plant, content to draw revenue from the leasing of existing Pullman cars.

Soon faced with eviction and under increasing economic duress, Pullman workers appealed for assistance to the American Railway Union and its leader, Eugene Debs, fresh from victory in the Great Northern Railroad strike. After debate within the organization, Debs agreed to help. George Pullman was offered an opportunity to negotiate a settlement of the dispute; when he answered in the negative, Debs called on ARU men to refuse to handle Pullman cars.

The boycott officially began on June 26, and the overwhelming response of railwaymen to his call surprised even Debs. In a few days, rail traffic west of Chicago came to a complete halt. Railwaymen not only refused to couple and detach Pullman cars; they also stopped trains which already carried them. By the beginning of July, the boycott spread eastward to New York.

A shutdown of the nation's railroads and commerce on an order even greater than in July 1877 was now threatened, and forces appeared to stem the revolt. The attorney general of the United States at the time, Richard Olney, assumed the lead. He advised lawyers for the railroads to seek injunctions against Debs and others for restraining trade (thus using the newly adopted Sherman Antitrust Act against labor) and for interfering with the passage of U.S. mails on trains. On July 2 a federal judge in Chicago issued an order with stiff penalties to halt the boycott. One day later, Olney convinced President Grover Cleveland of the necessity of sending federal troops to Chicago to enforce the injunction.

Events now marched to a harsh conclusion. Sporadic fighting broke out between ARU supporters and federal troops, with the toll eventually rising to twenty-five killed and sixty seriously wounded. A federal grand jury returned indictments against Debs and other union officials. A call then went out for a general strike, but the effort crumbled. Debs was arrested on July 17, and the Pullman Car Company reopened for business a day later with a new workforce; the ARU officially ended its boycott of Pullman cars in early August.

The Pullman strike and boycott had numerous repercussions. The federal judiciary now assumed an upper hand in labor disputes. Eugene Victor Debs went to jail and began his conversion to socialism, which would take him five times to national elections as the presidential candidate of the Socialist Party of America. The Pullman affair also effectively destroyed the ARU, leaving more conservative railway unions in command of the field. In fact, the defeat of the boycott and similar labor defeats at the time—most notably, the Homestead strike of 1892—convinced many trade union leaders that greater caution was in order, that objectives had to be narrowly defined now and large-scale uprisings avoided. The Pullman boycott and strike would be the last major nationally based labor disturbance of its kind for decades to come.

Strikes, however, remained a basic component of the American scene—they were now just more locally based. In fact, focusing on the great confrontations of the period obscures the very prevalence and range of strike activity in the last decades of the nineteenth century. Work stoppages occurred in great number in any given year, in all trades, and across America's industrial communities. Contemporary concern about the level of unrest actually allows for a fairly accurate accounting of strike actions during the period.

Starting in the late 1860s, trade unionists and social reformers called on state legislators to create state agencies responsible for gathering statistics on conditions of work, standards of living, and worker unrest. Collecting and disseminating information, it was hoped, would make the public aware of the deteriorating circumstances of working-class life and lead to legislative relief measures. Massachusetts led the way with the founding in 1870 of the Massachusetts Bureau of Statistics of Labor, and in the next three decades, twenty-nine other states followed suit. In 1885, the U.S. Congress similarly established a federal Bureau of Labor Statistics for the assembling of national data. From these various agencies issued surveys and reports on wages, hours of work, industrial accidents, family structure and means of support, patterns of consumption, housing conditions, union membership, and strike activity.

The statistics on strikes reveal a vast increase in job actions during the last decades of the nineteenth century. In 1881, across the United States, there occurred a recorded 474 work stoppages; in 1886, 1,432; and 1891, 1,717. Between 1881 and 1885, strikes averaged nearly 500 a year; the number rose to more than 1,000 on average in the latter part of the decade, and through the 1890s, there were on average 1,300 work stoppages each year. The numbers of workers involved and establishments affected also escalated. In the early 1880s, 125,000 workers typically participated in strikes each year; in the 1890s, 250,000. A high was reached in 1894, the year of the Pullman strike, when 505,000 Americans were counted as having joined a strike action. (It should be noted that these figures represent only 2 to 3 percent of the entire workforce at the time; however, the absolute numbers involved warranted alarm.) More than 12,000 businesses experienced work stoppages in the mid-1880s, with the figure rising to 30,000 in the 1890s.

No industry escaped the contagion. The railroads were the sites for the most dramatic confrontations of the period, yet rail strikes accounted for less than 2 percent of all work stoppages in the last decades of the century. Job actions of construction workers comprised the greatest number, 26 percent of the total. Coal miners were next, figuring in 10 percent of all strikes; in terms of numbers of workers involved, however, coal miners composed by far the largest battalion of strikers, 31 percent of the total number of workers engaged in work stoppages between 1880 and 1900. The greatest unrest in the coal industry was in the anthracite coal fields of eastern Pennsylvania. The Molly Maguires, a secret society of Irish miners, engaged in literal warfare with mine managers and their police forces there in the 1870s; more ordered job actions were conducted under the aegis of the United Mine Workers Union in the 1890s. With construction workers and miners forming a large proportion of the strikers of the period (and railroad workers an important if smaller part), an identification of strikers with male workers is easily made. But women workers also joined in the explosion of job stoppages. Clothing workers, a large percentage of whom were female, actually represented the third largest number of strikers during the period, 10 percent of the total.

The data collected by state and federal bureaus of labor statistics in the late nineteenth century also reveal that American workers walked off their jobs for a variety of reasons. They fought for wage increases and against wage reductions; for shorter hours, union recognition, controls on hiring, and union-determined work rules; and in sympathy with other strikers (nearly 10 percent of all job actions in the early 1890s were so-called sympathy strikes). Wages remained the key issue, however, figuring centrally in 40 to 60 percent of all strikes called in any given year during the period. Union recognition and rules regarding work assignments, discipline, and layoffs represented the next most important issues spurring stoppages. To gain their demands, workers stayed off their jobs typically for between fifteen and thirty days, and their perseverance often led to victory. The information gathered by labor bureaus indicates that strikers in the last decades of the century secured their demands in 47 percent of the total number of recorded work stoppages; they lost in 39 percent; and compromises were reached in 14 percent.

Despite the clear advantages held by business managers in countering the job actions of workers—access to the policing powers of government, the ability to hire strikebreakers in great numbers, especially with immigrants and African-Americans in desperate search of work—the deck obviously was not completely stacked against the strikers. One important weapon they had at their disposal was community support. An interesting statistic compiled by labor bureau officials speaks to this point. Strikes were recorded as ordered or not ordered by established unions. In the early 1880s, more than 50 percent of all strikes did not involve a formal trade union organization. The proportion of work stoppages orchestrated by unions rose over the next two decades, but by 1900, one-third of all strikes were still waged without union intervention.

The absolute grassroots insurgent nature of many late-nineteenth-century strikes has to be appreciated. State and federal investigative commissions established to determine the causes of the unrest of the period often searched in vain to find leaders or organizations to which responsibility could be assigned. The strikes also appeared to be as much community uprisings as work stoppages. Investigators found and local newspapers reported ample evidence of widespread support for strikers. Workers from trades not directly involved walked off their jobs in sympathy; local shopkeepers offered food and extended credit to families of strikers; editors of community newspapers blasted company officials for not dealing fairly with their employees; townsmen called up for service in state militia fraternized with their neighbors who were on strike and failed to take up positions in guarding business properties. Community members also took to the streets in protest with strikers. Arrest records for the period reveal people from all walks of life incarcerated for rioting, arson, and attacks on police forces during labor upheavals. Women participated in demonstrations as visibly as men. During railroad strikes, brigades of women greased and soaped tracks to impede the passage of trains.

Why did people who were not directly involved in labor disputes join in protest with striking workers in large numbers? Economic bad times is a contributing factor. Community uprisings accompanying strikes generally occurred during years of economic depression. Yet, material grievance alone cannot explain local insurgencies. The target of community aggression was corporate enterprise. During disputes, corporate property was attacked, not the businesses of local entrepreneurs. Tension exploded to riot when troops sent at the behest of corporate leaders entered the local scene. Reporters covering the disturbances discovered local shopkeepers sympathizing with striking workers and joining protest because the national based corporations threatened their existence; in the case of the railroads, proprietors were angered both at the physical incursions of the carriers and the seemingly unjust rates charged them for shipping goods. The very livelihoods and the autonomy of members of communities appeared challenged by the impersonal decisions made by executives in remote and unapproachable corporate headquarters.

The distrust of the first generation of Americans to be faced with the corporation is explicable; less easy to fathom is the violence. Protestors picked up bricks and rifles to defend their communities and republican ideals. The common ownership of guns in a nation where the right to bear arms was constitutionally protected is part of the explanation for the dramatic loss of life and limb in labor uprisings. The frontier also played somewhat of a role. Not every American community witnessed death and destruction during the labor upheavals. However, historians have been unable to discern patterns of insurgency; labor insurrection occurred in both metropolitan areas and small towns. Unrest unfolded in medium-sized cities as well, but notably in newly developed industrial communities—common to the Midwest—without established elites or ways. Finally, and crucially, the determination of American corporate executives to suppress strikes and unions at all cost and to employ public and private police forces to silence protest was also a key ingredient in the remarkably fierce battles that transpired. Fire was fought with fire.

The Role of Unions in Labor Unrest

Spontaneity and community support notably marked labor protest, but over the course of the last decades of the century, trade unions assumed a greater presence and importance in strikes. A trade union revival had first occurred in the 1850s. With economic recovery after the depression of the late 1830s and early 1840s, skilled workers on the local level reestablished their antebellum craft societies and then joined with workers from other localities to found national organizations. The arrival of skilled German and British workers, who were highly politicized and had trade union experience, fueled the rebirth; and vast improvements in transportation and communications during the 1850s allowed for (and demanded) greater cooperation across geographical boundaries.

The revival of trade unionism in the 1850s was followed by increased labor organization and strikes during the Civil War, especially as workers attempted to keep their wages in line with rising wartime prices. In another kind of labor protest, white workers in the North during the war took to the streets to protest the inequities of the military draft system, taking their wrath out on recruitment officers and, tragically, often on African-Americans as well.

An expansion of trade union activity and membership during the Civil War led to the formation of the National Labor Union in 1866, the nation's first national federation of unions. Under the leadership of William Sylvis, an articulate iron molder, the NLU convened yearly conventions where trade unionists and various labor sympathizers discussed issues of the day. Delegates supported motions in favor of the eight-hour day, government monetary policies that favored debtors, federal land distribution programs for working people, increased efforts at labor organizing, and most notably, the building of so-called producers' cooperatives (if yeoman producership was increasingly untenable with rapid industrialization, then republican principles could be sustained with yeoman cooperatives). The NLU maintained a visible presence for five years, with some strike and legislative success, but the movement then dissipated and passed from the scene. The organization failed to survive the depression of 1873 and was further incapacitated by internal divisions over economic policy, political activism, and gender and racial issues. From the ashes of the NLU, however, rose another trade union federation, drawing upon and extending many of its predecessor's ideals, but having a much greater impact. This organization was the Knights of Labor.

Uriah Stephens and James Wright, skilled garment cutters, founded the Noble and Holy Order of the Knights of Labor in Philadelphia in 1869 as a secret organization. Little is known about the society in its early years, though it did survive the depression of 1873 and included some 500 lodges and 6,000 members by 1877. In 1878, the order went public (although a good many of the rituals of the original clandestine organization were maintained), and under the leadership of Terrence Powderly, a machinist by trade, members of the "producing classes," regardless of occupation, nationality, race, religion, or sex, were encouraged to join a movement bent on achieving better working conditions and a new social order based on equality and cooperation. Explicitly barred from membership were bankers, stockbrokers, lawyers, liquor dealers (temperance remained a guiding principle of the organization), and gamblers.

In the early 1880s, the society grew and spread throughout the country; recent analyses of the organization's surviving documents reveal lodges established in every state and county of the nation. In 1885, striking Knights railwaymen in the West won their spectacular victory over the powerful Jay Gould, boosting the organization's fortunes and membership to 750,000. The year 1886 saw the order at its peak of prominence as Knights of Labor members rushed into political activity, launching successful independent party ventures throughout the United States. From this zenith of visibility and impact, however, the Knights of Labor declined precipitously under the force of increased employer resistance to Knights-led strikes, internal divisions, and the defection of craft unionists from the cause. By the mid-1890s, few traces of the order remained.

An overview of the history of the Knights can thus be rendered, but understanding the Knights phenomenon has always provided difficulties for historians. For one thing, the activities of the federation varied from year to year and place to place. The organization itself was a crazy quilt of local assemblies of individuals, neighborhood groups, reform associations, existing craft unions, and workers variously organized by factory, trade, and geographical locale, and wider city, district, and state assemblies. The order stood for the enrollment of all workers—upholding the principle of so-called industrial unionism—yet craft unions joined the federation and maintained their autonomy and identities. The order made unprecedented strides in organizing women and African-American workers, though gender and racial divisions still marked its history. Knights activists moved into independent political party activity with a vengeance in 1886, yet an ambivalence toward politics, politicians, and the role of government—stemming from the republican and antistatist sentiments and ideals of those drawn into the movement—blunted the Knights' political initiatives once Knights candidates found themselves actually elected to office. Temperance and chivalric behavior were hallmarks of the order, yet a number of Knights officials were involved in swindles and intrigues that rivaled the worst corruptions of the age. The organization stood officially for the abandonment of the wage labor system, but with the exception of support for public ownership of financial and transportation institutions, Knights leaders openly repudiated socialism. Knights officials upheld the building of producers' cooperatives as a principal goal and ideal, yet their actual record on establishing cooperative ventures was poor. Finally, and most notably, while the organization's leadership formally renounced strikes and espoused harmonious relations with fair-minded, hard-working employers and arbitration in the case of disputes, its members nonetheless participated fervently in hundreds of strikes under the Knights banner.

The Knights of Labor thus defies simple characterization. Can it be deemed a backward-looking movement aimed at recreating a mythical yeoman producer past, as some historians have suggested? Not really. Knights leaders and followers harked back to a more harmonious time, yet they were permanent wage laborers, decidedly immersed in modern issues like arbitration proceedings, challenging traditional mainstream politics, and struggling to find new forms of relationships. Was the organization instead a typical reform movement merely bent on extending the benefits of capitalism to the yet unbenefitted? Not really, again. The Knights were ambivalent about the capitalist system, definitely antagonistic to corporate or monopoly capitalism, and certainly not boosters of capitalism in general.

Was the organization an alternative movement looking forward to the building of a "cooperative commonwealth" (a frequently invoked phrase) to replace the bureaucratic and corporate future that loomed ahead? There is evidence to substantiate this interpretation, but how widely this vision was held by those enrolled is unclear. Was it, as other historians have suggested, a democratic movement at heart, an attempt to significantly widen political participation. Ample proof exists for this contention as well, though the notion hardly provides a comprehensive understanding. Was it simply a trade union effort? Certainly not.

Perhaps the only solution is to term the Knights of Labor an amorphous social movement of laboring people; the extent to which hundreds of thousands of American working men and women were enlisted, educated, and politicized in the 1870s and 1880s is what is to be ultimately appreciated. Recognition of the mobilization is apparent from studies of local Knights' activity and the countless meetings, lectures, parades, and picnics attended by those moved to join. If historians have to remain unsure in their estimations of the Knights, the Knights of Labor moment in American history nonetheless continues to fascinate.

Some trade unionists of the era, however, viewed the Knights venture as just pure folly. Leaders of the so-called brotherhoods of railway workers, for example, expressly forbade their members from joining the Knights; the brotherhoods, in fact, avoided participation in all of the great uprisings of the day. Leaders of most craft unions similarly viewed the Knights with a jaundiced eye. Among detractors of the Knights, the key and historically critical figure was Samuel Gompers.

Samuel Gompers was born in England in 1850 and immigrated to New York City at the age of thirteen. Entering the cigar-making trade, he found himself immersed in a community of skilled English and German immigrant workers who ate, drank, and breathed Marxism and socialism. Gompers thus received lessons about the seeming hard truths of capitalism: a permanent proletariat had been formed; capital was concentrating and gaining overwhelming economic and political power; realism was in order; focused and well-organized trade unions had to be established to achieve gains for workers.

Despite this sober perspective, Gompers and his fellow craft trade unionists were swept into the Knights crusade. Having risen to the leadership of the Cigar Makers' International Union, he joined with other skilled men in 1881 in founding the Federation of Organized Trades and Labor Unions, which allied with the Knights. Over the next five years, however, Gompers and his associates became increasingly disenchanted with the movement. They resented Terrence Powderly's attempt to dictate policy, found themselves engaged in various jurisdictional disputes (what constituted a local or an assembly of the Knights organization remained a fuzzy and disputed matter), deemed most Knights pronouncements as naive and Knights-sponsored strikes as quixotic, and saw the sudden move of Knights members into independent party politics as misguided and wasteful of valuable time and energy. In 1886, Gompers thus led numerous craft unions out of the Knights of Labor and into his newly formed American Federation of Labor. The absolute loss in membership and the defection of the best-organized of the Knights' members contributed greatly to that order's subsequent demise.

Under Samuel Gompers' leadership, both the constituent unions of the AFL and the offices of the federation grew in strength and permanence. Gompers himself became the most important labor leader in the country, listened to by politicians, businessmen, and the press. Until his death in 1924, he was arguably the chief spokesman for the working people of the United States. This all occurred, however, because Gompers forged a limited agenda.

Gompers' strategy for the AFL including the following:

1. Organize the organizable—that is, skilled workers who possessed leverage in the workplace and who could win victories.

2. Do not pour energies into enrolling less skilled, easily replaced workers. In practice, this meant ignoring women, blacks and most immigrants.

3. Build strong organizations with high dues, well-paid officers, and strike funds and other benefits that would engender the great loyalty of members.

4. Fight for what can be gained—"bread and butter" issues—meaning higher wages within reason, shorter hours, and work rules that curtailed the arbitrary decision making of firm managers.

5. Do not become involved and waste time and money in independent political action, much less radical political activity. Work within the system, in other words.

6. Support mainstream politicians of any party who favor prolabor legislation.

Gompers' strategy proved successful—in many ways it was programmed to be—and with each victory for AFL men, Gompers could upbraid his many detractors in the labor movement and point to the wisdom of his stance.

The history of Samuel Gompers and the AFL in the late nineteenth century, like that of the Knights of Labor, is not without its inconsistencies. Gompers developed his strategy over time and with experience, not presciently or at once. He fashioned his approach to trade unionism after his experiences with the Knights, through watching corporate capital become entrenched and powerful, after disastrous defeats for labor at Homestead and Pullman, in light of increasing judicial assaults on trade union activity and conservative court rulings against the government regulation of working conditions, after turning back significant challenges to his leadership by militants from within the AFL, and after seeing the efficacy of ingratiating himself to business leaders (as he warned, if they did not deal with him and the skilled workers he represented, they would have to deal with more radical groups within the laboring community). Gompers also asserted himself as a visible spokesman, but he insisted that the AFL be as decentralized as possible and that constituent unions be afforded maximum autonomy (the example of Powderly glared in his mind). Gompers also upheld the principle of craft unionism, yet the AFL included so-called federal unions, such as the United Mine Workers, which organized workers across skill levels. While in practice the AFL rendered little assistance to women and African-American workers who struggled to form unions, Gompers himself spoke against racial and sexual prejudice, opposed segregated unions, and warned that without outreach, strikes would be lost through the employment of strikebreakers who remained antagonistic to the exclusive craft unions. Gompers similarly argued for labor to be nonpartisan and nonreliant on politicians (he espoused "voluntarism"), yet he would ultimately hitch the fortunes of the AFL to the Democratic Party. Finally, Gompers placed a definite conservative stamp on the American trade union movement, yet as noted, he constantly faced challenges from groups within the AFL who mobilized to see the federation adopt a much more militant, socialistic, and inclusive posture.

There are a number of great ironies in the story of the Knights of Labor, the American Federation of Labor, and other labor campaigns in the last decades of the nineteenth century. Samuel Gompers, schooled in Marxism, became a force for moderation and acceptance of the powers that be. Eugene Victor Debs, on the other hand, born in small-town America, steeped in American republican traditions, would emerge a socialist after defeat at the hands of the powerful Pullman Company and the U.S. government. In launching effective strikes Gompers and his associates challenged capital head on, but not the capitalist system. The Knights, conversely, in raising on high the notion of a cooperative commonwealth and in their spontaneous protests, challenged the capitalist order but not, effectively, capital.

Egging on labor revolt during the late nineteenth century were also a host of socialists, anarchists, and other radicals, operating in such organizations as the Socialist Labor Party. During the great railroad strikes of July 1877, for example, leaders of the socialist Workingmen's Party assumed a key role in mobilizing protest in the city of St. Louis, one of the only instances during the nationwide stoppage where there is clear evidence of leadership and organization. Individuals and groups—whether craft unionists, Knights, or radicals—thus contributed to the labor upheavals of the period. Ultimately, however, a lesser role has to be assigned to all of them, for the unrest is only partially explicable in terms of formal organization and agitation. General resentment among working people and often among their middle-class neighbors against the emerging power of the corporation played a larger role. So, too, did anger about economic hard times.

Economic Conditions and Labor Unrest

The economy provides an explanation for the labor upheavals of the time, although neither a complete nor a straightforward answer. The American economy in the late nineteenth century is a study in contrasts and complexity. The last three decades of the century witnessed extraordinary growth in industrial (and agricultural) production, the creation of an American industrial heartland, the rise of the corporation and new ways of conducting business, and the appearance of startling new technologies: few could imagine in 1875 the world of telephones, electrical lighting, steel-beamed skyscrapers, and even cars with gasoline engines that had emerged by 1900. Yet, during that same time span Americans experienced two calamitous depressions, one in 1873–78 and another in 1893–97, and a serious recession, from 1884 to 1886. These particularly bad times not only saw monumental strikes, but also demonstrations in American cities by the unemployed—who represented upwards of 30 percent of the workforce—for public relief programs and work projects; during the depression of 1893 thousands of the dispossessed joined Coxey's Army, a notable march led by one Joseph Coxey across the country to Washington, D.C., in search of federal help. But economic ebbing and difficulties did not just mark a few crisis years. Economic historians have measured declines in prices, profits, nominal interest rates, per capita output, and rates of growth across the entire last three decades of the century.

The annual rate of growth in the country's gross national product stood at 6.5 percent in the early 1870s and 3.6 percent in the 1890s. Output throughout the period increased as America became the ascendent industrial nation in the world, but the declining *rates* of growth have puzzled scholars and there is little consensus as to the explanation. The boom-and-bust character of the economy in the late nineteenth century has been pictured by some as just part and parcel of business cycles normal to the capitalist system. According to this view, producers reacting to rising prices increase their output until a point is reached where supply outruns effective demand and a retrenchment ensues; when supplies subsequently dwindle and demand increases, a new cycle of growth begins. The problem with picturing the American economy of late nineteenth century as just exhibiting normal business fluctuations is that the crises of the period were hardly normal downswings; the successive declines in annual growth rates also remain unexplained.

Other historians have tried to understand declines in the period as part of longer swings in economic activity. These scholars see long periods of growth stimulated by some prime development or mover, such as the railroads or western expansion, and an eventual extended exhaustion of initiative and dissolution until there occurs another upswing. In this scenario, the ending of railroad building or the closing of the frontier led to a general stagnation in the late nineteenth century; electrification or some other new development would then produce a new upswing. At a very high level of generalization, the long-swing approach to understanding the economic record of the nineteenth century is persuasive, but a great deal of history is wiped away with this sweeping perspective.

Other historians have emphasized particular events or circumstances in accounting for the vicissitudes of the economy in the late nineteenth century. The speculative mania of the times has been noted. Both major depressions of the period occurred with the failure of key investment schemes and brokerage houses; the collapse of the securities market in each instance erased confidence and incentives for production. Severe competition generated by transportation and communications improvements similarly can explain the falling prices and profits of the period. Other historians have pointed to poor purchasing power as an answer for decline, while a counterargument places the responsibility for declining rates of growth on high wages. Real wages did, in fact, rise more than 70 percent between 1865 and 1890 because of rapidly falling prices and the agitation of workers. For some economic historians, high rates of savings or investment combined with no increase in labor productivity and rising real wages led to a profit squeeze and falling rates of growth. The determined efforts of corporate managers to wrest control of production from skilled workers and to reduce wages are all the more understandable from this viewpoint, which locates class conflict and the basic tensions between capital and labor in the economic problems of the times.

Opposing perspectives on the American economy in the late nineteenth century (and the specific arguments of scholars themselves in many instances) can provide for great confusion. The tangle of positions is in some sense unavoidable, for the period was marked by great disjunctures and, more to the point, by both extraordinary expansion and extraordinary crisis. The labor upheavals of the era, in this regard, cannot simply be laid to the economy because there was nothing simple about the economy. This is reiterated when moving from the economy in general to the level of individuals and the actual economic wherewithal and prospects of workers and their families.

The late nineteenth century was marked by enormous disparities in wealth and income, even among workers. A great gap emerged between the top and the bottom in the society. In 1890, the top 1 percent of wealth holders owned 51 percent of all property, the bottom 44 percent just 1.2 percent. In this top 1 percent were 125,000 families, and they, on average, owned $264,000 of real and personal property. The bottom 44 percent, on the other hand, comprised 5.5 million families, and they owned, on average, $150 of property. If the ranks of the wealthy are widened somewhat to include the top 12 percent of families, the figures reveal this top group to be holding 86 percent of the wealth, which means that the other 88 percent of the population, including the poor and people of middling status, held just 14 percent of all wealth. The wealthiest Americans could afford to live with unprecedented extravagance. Their opulent lifestyles, at odds with old hearty republican values, bred both critique and cartooning from the press and resentment among the general population (another ingredient in the labor upheavals of the day).

Workers did make gains during the period—falling prices and their own organizing led to advances in real income—but the progress was not widely shared. Skilled workers, for example, earned between $700 and $900 a year, just $200–$300 less than highly paid clerks and professionals. Well-paid skilled men could afford homes with furnishings that made for respectability. They could also afford to keep their older children out of the labor force. Sons and daughters of the "labor aristocracy" began enrolling in new high school commercial course degree programs, receiving degrees that allowed them entrance into the burgeoning world of white-collar employment. Clerical jobs were not necessarily better-paying than skilled trades positions, but they represented more desirable work, especially for young women who wished to avoid domestic service and factory positions. While there was growing anger over the corporation, a pioneer generation of Americans soon found a place in the corporate world handling the massive sales, accounting, and paper work of these new entities.

For the great majority of working people, however, the late nineteenth century was a time of just making ends meet or of living in absolute poverty. Most factory workers earned between $400 and $500 a year, and the least skilled, little more than $250. Families at this level could only afford to rent and reside in the squalid tenement buildings that emerged in the period as a feature of many industrial cities. The earnings of older children were essential for family survival. Family budget studies of the period reveal that children of poorer working-class families collectively contributed to more than 40 percent of family income. These youngsters could not attend school or did so on an intermittent basis.

Working-class families also depended vitally on the largely unpaid labor of wives and mothers. Only a small percentage of white married women—between 3 and 5 percent—entered the labor force during the last decades of the nineteenth century. They did so in extreme circumstances, after the desertion, injury, or death (the latter two often from accidents on the job) of the male heads of households. The great majority of white working-class women remained at home, spending their days endlessly sewing and mending clothes, cleaning, laundering, fetching water, shopping, cooking, attending to child care, managing the taking in of boarders, laboring on piecework, and hawking and scavenging in the street. Historians have affixed a dollar value to these tasks and discovered that the sum of goods and services provided by mothers was twice as great as they could have earned on average in paid employment outside of the home. African-American women do provide an exception here; with black men in northern areas relegated to the lowest paid and most irregular of jobs, married black women entered the labor force in great numbers, with more than 30 percent in the late nineteenth century earning wages, largely in domestic service and commercial laundries.

The material circumstances of American workers in the late nineteenth century thus varied widely. Cross-national comparisons reveal that skilled men in the United States achieved standards of living far more comfortable than their direct counterparts in other countries at the time;

for the lesser skilled, conditions in the United States were only slightly improved. American workers, however, no matter what their means, faced a common and serious problem throughout the period, namely, highly irregular and uncertain employment.

The last decades of the nineteenth century witnessed moments of massive unemployment, but even in better times, few laboring people worked a full stint of days during the course of the year. Seasonal changes in demand and constantly fluctuating business circumstances meant that employers frequently closed shop and furloughed their employees for months at a time. A federal study of the railroad industry in 1889, for example, revealed that due to irregular commerce only one-fourth of all railway employees in the country worked a full complement of days; near 60 percent were employed and paid for what amounted to a half year's work. Similarly, statistics for the state of Massachusetts indicate that upwards of 30 percent of all workers could expect to be out of work at some point during the working year. The irregular employment and incomes of heads of households made attention to budgeting an imperative as well as the early entrance of older children into the labor force.

The corporation actually heightened levels of uncertainty, particularly for production workers. While white-collar employees in corporations could find relatively stable employment, manual workers found themselves at the mercy of plant supervisors and foremen, who controlled hiring, training, work assignments, discipline, and most important, the matter of what would happen in terms of continued employment in the cases of accidents and slack business conditions. The capricious rule of supervisors, in fact, was a major issue in the labor upheavals of the day. The testimony of Franklin Mills, a railway employee discharged by the Baltimore & Ohio Railroad Company for his participation in the Pullman boycott of 1894, to a special federal investigating commission investigating the upheaval is revealing on this score:

Commissioner Kernan: What was the feeling among the men on the Baltimore & Ohio with regard to striking prior to the time they struck?

Mills: It was not very favorable.

Commissioner Kernan: Had there been any cuts in wages about which they were dissatisfied?

Mills: Not lately. The most of the difficulty on the Baltimore & Ohio was favoritism, pets and maladministration of some of the petty officers.

To counter the whim of supervisors, workers in the late nineteenth century organized and attempted to gain union contracts that included work rules aimed at a modicum of justice and security on the shop floor.

Economics thus played a role in the labor unrest of the late nineteenth century. But it was not a matter of material hardship per se; the material circumstances of workers varied. It was critically a matter of economic security. The upheavals, however, were also about power, at a time of the rise of the corporation: about personal power and the ability to maintain control over one's life, and importantly, about public power as well, and how Americans would relate to one another and function as members of communities and a nation at large.

Farmers' Protests

Other voices of protest were also raised in American communities during the last decades of the nineteenth century. American farmers, most notably, engaged in active organization throughout the period. In the late 1860s they first formed local branches of the new Patrons of Husbandry, or the Grange. The Grange began as a nationwide association of social and educational clubs, but in the 1870s members of the Grange began forming purchasing and marketing cooperatives and lobbied at the state level for legislation regulating the pricing policies of railroads and grain elevator operators. So-called Grange laws outlawed pooling arrangements and discriminatory rate practices that favored certain shippers over others. The Grange laws represented a historical first step toward greater government regulation of business in the United States. When the Supreme Court declared that state regulatory legislation could not apply to enterprises operating across state boundaries, farmers and other shippers joined forces to see passage in the U.S. Congress in 1887 of the Interstate Commerce Commission Act. The bill established the Interstate Commerce Commission, the nation's first federal regulatory agency, specifically to prevent collusive activity among railroad companies. Although afforded a vague mandate and powers, the ICC marked a new departure in the federal government's role in economic affairs.

American farmers also assumed a key place in a fractious debate of the times involving the currency of the nation. The question of what constituted a proper monetary system for a democratic republic had occupied Americans since the 1790s and came to the fore during the antebellum period with repeated discussions on the creation of a national bank. Opponents of the bank had argued that managers of a central bank would possess dangerous powers and that, in exerting strict controls over the currency, their policies would inevitably favor creditors at the expense of debt-ridden farmers and workers.

Monetary matters became especially heated during and after the Civil

War. To pay for the war, the federal government had issued so-called greenbacks, paper notes not backed by or redeemable in precious metals. The greenbacks fueled a drastic wartime inflation, and fiscal conservatives issued an immediate call to have them withdrawn as a circulating medium. Labor and farm groups countered with petitions to retain the greenbacks because for them a loose money system meant eased access to credit, an opportunity to pay back debts in depreciated currency, and a chance to reverse a deflationary cycle that kept farm prices and wages low. Political defeats in the early 1870s, when Congress moved both to place the country on a strict gold standard and to withdraw the greenbacks, pushed farm and labor adherents momentarily into independent political party activity.

In support of liberal monetary policies, they formed the Greenback Party in the mid-1870s. Advocacy of a loose paper money currency actually represented a reversal on the part of labor activists; during the antebellum period, labor leaders had championed a metal-coin-based system because workers were often paid in scrip, notes that were to be used in purchasing goods at company stores but that often proved worthless. Greenbackers did not succeed in convincing Congress to sustain a loose paper currency. However, pressure by farmers and workers did lead to legislation that allowed for increased federal government purchases of silver; in greater supply and cheaper than gold, silver served as a base for the release of greater amounts of redeemable paper notes.

Some scholars have considered the fixation of farmers and others on monetary issues in the late nineteenth century an irrational obsession, one that blinded them to the reality that farming had become irreversibly more capital-intensive and concentrated in fewer hands. Dialogue on monetary matters in the 1870s and 1880s, however, was not fanciful. The question of the money supply was tied to a basic concern for the shape of the republic and whether American laboring people—whether working on the land or in shops and factories—could remain economically and politically independent. Supporters of liberal monetary practices entered into debate with serious intentions and judicious arguments. For farmers, the question of the money supply also became wrapped up in another issue, and that involved the farm cooperatives that they strived to build during the period. In the 1880s, farm protest reached a new phase with the emergence of the Farmers' Alliance, an organization spurring the formation of local farmers' purchasing and marketing cooperatives. Farmers mobilized to buy seed and other supplies in volume at wholesale prices and invest together in expensive farm machinery. Rather than compete with each other, they moved also to pool their harvests and market in concert, waiting for the best prices. The cooperative move-

ment headed by the Farmers' Alliance quickly became a crusade, enrolling, educating, and politicizing farmers.

The farm co-ops founded in the 1880s needed eased access to credit,
and the issue of the money supply thereby became linked to the cause of
cooperativism. A key reform pushed by leaders of the Farmers' Alliance
during the decade was the so-called subtreasury plan, an ambitious scheme
that called for the federal government to establish a nationwide system of
agricultural warehouses. Under the program individual farmers or farmers organized into cooperatives would store their crops in government
storage facilities, avoiding the perceived exorbitant fees charged by private operators; and in turn, their crops would serve as collateral for the
issuance of low-interest government loans. Supporters of farm cooperatives hinged the success of their movement on the enactment of the subtreasury plan.

When politicians from the two major parties turned deaf ears to calls
for enacting such a plan, farm protest entered a third stage as farmers in
the Alliance movement turned aggressively to independent party politics
at the local level. After extraordinary success in 1890 in electing local
officials and representatives, Alliance supporters then moved to found a
national third party. At a convention on July 4, 1892, they formed the
People's or Populist Party and drew up a platform that called for enactment of the subtreasury plan, an expanded monetary system based on
silver, public ownership of the nation's railroads and telegraphs, and
other political reforms aimed at increased accountability of politicians
and participation of citizens in political decision making. In November
1892 the Populist Party's candidate for president, James B. Weaver, received 8.5 percent of the total vote and the electoral college votes of six
states, the most successful insurgent political effort up to that point in
American history.

From this high point, the Populist effort dissipated. The farmers' party
was not able to take advantage of popular discontent spawned by the
depression of 1893; subsequent splits among supporters sapped the movement; decisions to concentrate campaigns on the silver issue blunted its
appeal; and the move itself toward official politics dampened the grass-
roots engagement that had marked farm protest from the Grange through
the Alliance. Farm protest in general receded from the American scene
after 1897. Rising demand for farm products accompanying European
crop failures and the massive turn-of-the century immigration of southern and eastern European peoples to northern industrial centers of the
United States, in conjunction with an increase in the money supply afforded, ironically, by notable discoveries of gold in Alaska and South
Africa, sparked an inflationary cycle that boosted farm prices and in-

comes and ushered in a twenty-year period of welcome prosperity for the American farm community. The farmers' movement thus ebbed, but the ideas and sentiments that had fueled activism for three decades did not disappear and would influence mainstream politics for decades.

The protests of workers and farmers in the late nineteenth century invite comparisons. Farmers and workers in fact lent support to each other at various points, and their movements shared many features. As with labor activism, the strength and meaning of farm protest is best appreciated at the local level and is obscured by a focus on organizations and leaders. The farmers' movement also had a similarly varied cast, with the content and form of protest varying between regions and over time. Not coincidentally, scholars have offered vying interpretations of the movement, categorizing farmer activism as simple interest-group politics (that is, as a straightforward effort by a beleaguered segment of the population to receive some relief during particularly bad times); as a backward-looking cultural movement (in other words, a misguided endeavor to return the nation to a bygone, and mythical, age of yeoman producership); as a radical critique of the new bureaucratic, corporatist order and an attempt to forge a cooperativist alternative; and as a struggle aimed at heart at the widening of political participation. There is truth in all of the above interpretations, and as with labor mobilization, perhaps the crucial point is precisely the protean character of the farm protest movement in the late nineteenth century. Finally, for farmers and labor activists of the period, economic well-being and equal citizenship (for white males) were one and the same issue.

The farmers' and workers' movements also diverged in significant ways. Both upheld the ideal of the cooperative commonwealth, but in the building of producers' cooperatives, the farmers' record was far more impressive. The farmers also held much less ambivalent views on the state and state power. They proposed numerous government programs and were key agents in moving American public opinion toward acceptance of government involvement in economic affairs. Farm activists thus had a greater impact on politics than their counterparts in the labor movement. Finally, a tension existed. Farmers ostensibly were property holders and potential employers of labor; their class interests seemingly remained apart from workers. Shared republican and producer values and the very reality that growing proportions of farmers were tenants and land workers and not land holders kept farm and labor activists in alliance. Knights of Labor members and Alliance farmers and Populists thus cooperated in each other's campaigns; the more class-conscious and skeptical Samuel Gompers, not surprisingly, kept his distance from the farm protest.

Other Voices of Protest

The farmer and labor unrest that fed on and contributed to the turbulence of the times was accompanied by a literature of protest that also was both a product and agent of change. Hundreds of thousands of Americans, for example, read and were moved by *Progress and Poverty* by Henry George, a book published in 1879. They may not have fully understood George's so-called single-tax theory, but his pleas to curb large-scale land speculation struck a chord in Americans who still believed that the rewards of the society were due solely to those who toiled. Supporters of George rallied to his strong but losing bid for the mayor's office of New York City in 1886, a crusade conducted under the banner of the United Labor Party. Americans in great numbers similarly read Henry Demarest Lloyd's *Wealth against Commonwealth,* a blistering study of and attack on corporate power. A corps of investigative journalists would soon follow Lloyd's lead with popular newspaper exposés of the underhanded practices of the robber barons and their allies. Ruthless tycoons also appeared in a spate of new, highly realistic novels as novelists such as Frank Norris and Theodore Dreiser joined newspaper reporters in presenting the public with damning critiques of business activity. Newspaper photographers also created vivid portraits of the exploitative working and squalid living conditions now endured by America's laboring people.

A remarkable group of women contributed to the literature of protest. Among the first generation of college-educated women and often religiously inspired, these emerging agitators left the comforts of their middle-class homes to move into the teeming working-class, growing immigrant wards of the nation's industrial cities to establish settlement houses, community centers offering neighborhood residents medical, legal, vocational, and recreational services. In the late nineteenth century, these women used the social settlements as bases to launch exhaustive surveys of working and living conditions in the areas they served. Their published reports, backed by their example of commitment, also helped shift popular opinion. They forced Americans to consider the social rather than personal causes of poverty and the necessity of public programs of relief. Female so-called social justice activists, such as Jane Addams and Florence Kelley, later played leading roles in lobbying for and shaping the nation's first government welfare programs, despite the continued denial of the franchise and political office to women.

Settlement house workers found allies for their cause in a new generation of academics and clergymen. In the 1880s, American universities first established departments for the study of economics, politics, and other social sciences. An influential cohort of professionally trained students of American society thus appeared. These scholars also conducted surveys of contemporary urban and industrial conditions, but they played a greater part in persuading Americans to think of themselves as part of an interactive whole and, relatedly, of the need for "social" policies. A group of inspired Protestant ministers delivered the same message from their pulpits. They formed the so-called Social Gospel movement and were strong advocates of concerted programs of care and assistance to the unfortunate.

Social Gospel clergy and other socially minded writers of the late nineteenth century wrote out of concern for the downtrodden and the changing circumstances of American life, but also to counter a rival viewpoint of the age espoused by leading businessmen of the period and other conservative thinkers. These conservatives penned a different creed. Borrowing metaphors with great license from Charles Darwin's newly read theories of evolution, they pictured existence as a pitched battle among individuals, with progress affordable only through survival of the strongest and fittest. They warned against naive interventions that circumvented the natural order of things. A laissez-faire literature and position thus emerged in the late nineteenth century, but a credo that must be distinguished from an earlier individualist ethos. Business figures constructed their arguments to justify their great accumulations of wealth and the powers held by the large-scale organizations they headed. They wrote, too, in reaction to rising calls for public relief programs and the regulation of their business practices. There is perhaps no irony then that they appeared as great advocates of individualism and free competition at a time when they were acting increasingly in associational ways and building great bureaucracies.

The notion of individualism as manifest in late-nineteenth-century laissez-faire practice would have appeared foreign to Americans of a prior time. For them, individualism was wrapped up in the whole question of the creation and maintenance of a democratic republic. Independence, competency, virtue, and citizenship were key values. Laissez-faire thought was in some sense a purely political tactic as well, a rhetoric to be invoked by business leaders at particular moments for particular advantage. Corporate leaders and people of standing had always supported public initiatives aimed at economic development and social order, and in the late nineteenth century the state loomed for them even more crucially as a means to achieve economic and political stability.

The literature of protest of the last decades of the nineteenth century as well as conservative creeds of the times painted a most somber portrait of America. Alongside the highly realistic writing of the age emerged an equally influential utopian literature. Utopian novels were especially popular in the 1880s and early 1890s, when adherents of the Knights of Labor and the Farmers' Alliance were also searching for alternatives to contemporary developments. Utopianists offered their readers visions of a bountiful, egalitarian, harmonious, and, above all, well-ordered future for America. Edward Bellamy's *Looking Backward*, published in 1888, was by far the most popular of the genre. Bellamy actually presented a highly technocratic and regimented future society, an ideal at odds with the more self-regulated and cooperative commonwealth imagined by visionaries in the labor and farmers' movements. Yet, thousands of Americans joined so-called Nationalist Clubs established to promote Bellamy's dream. The thought of a more administered political economic order was definitely in the air.

Economic crisis upon crisis, the challenge that the corporation posed to Americans, the dislocations produced by massive immigration and urbanization—all demanded response; the protests of workers and farmers in the late nineteenth century and the dissenting literature of social commentators and activists forced the action. Americans at the turn of the new century would begin to reconstitute their society. A number of broad points can be made about this process to place the industrializing of America in the nineteenth century in sharper perspective.

First, the creation of a new political economic order occurred slowly, in fits and starts, and over a good fifty-year period. Historians have tended to foreshorten this story, seeing America critically reconstructed during the so-called Progressive Era of the first decades of the twentieth century or during the New Deal period of the 1930s and 1940s. In fact, the process began in the last decades of the nineteenth century and continued through the Second World War.

Second, the new political economic order defies easy labelling. At the mid-twentieth century the United States was marked by the following features: corporate dominance in the economy; mass production and a mass consumer society; bureaucratic procedures and institutions; accommodations between capital and labor; government patterning of economic growth and spending to sustain high levels of demand; state welfare policies to place a floor on social suffering and maintain purchasing power; arms production as a foundation block of the economy; and international economic, diplomatic, and military power. America by the late 1940s bore no resemblance to the United States of July 1877, when the American people stepped so unknowingly into the future. No distinct title fits this new entity.

Third, the process was uneven and incomplete. Small and medium-sized firms persisted, though operating now at the periphery of the corporate core of the economy. Conflict endured on the American shop floor. The federal system, states' rights traditions, southern political intransigence, and free enterprise ideology continued to block attempts at a fully administered economic and social welfare order. Certain groups, and particularly African-Americans, remained outsiders.

Fourth, the reconstitution of America was not orchestrated, nor did it involve a single set of actors. The process is best visualized as the slow, interrupted building of a house. Different groups of Americans added bricks at different places and moments. A whole edifice emerged because each of the parties to the construction sought the stabilizing of economic, social, and political life.

Fifth, the groups involved acted in private and public ways. For example, businessmen in the corporate sphere first looked among themselves to solve problems of overcompetition; they reached informal and formal accords to control prices and merged as a measure of last resort. They also met to standardize technical practices. (Railroad managers typically led the way here in the late nineteenth century by collectively adopting standard gauge track, lading procedures, and time zones.) Businessmen also increasingly looked within their own firms for solutions—opting for product line diversification, decentralized management, and even benevolent programs that engendered greater loyalty from their workers.

But leaders of nationally based enterprises also sought the services of government to bring more certain business conditions. They petitioned government officials to help them secure access to foreign markets; the National Association of Manufacturers was established in 1895 expressly for that purpose. They relied on a compliant federal judiciary that established corporate rights. They especially used the state to curb labor unrest; between 1877 and 1910, troops were marshalled in more than 500 instances to quell strikes. Corporate executives also supported government regulations that allowed for a public seal of approval on mergers to preempt antitrust suits; government licensing and rate setting that could help curb competition; and even welfare programs—such as workmen's compensation—that could further rationalize their own businesses and cast them in a less scurrilous light publicly. Finally, they pressed for government orders. As early as the 1890s, major companies in the steel industry began to rely heavily on purchases of armor plate by the U.S. Navy for battleship construction. As striking steelworkers clashed with Pinkerton guards at the great Homestead strike of 1892, naval shipbuilding in fact ground to a halt hundreds of miles away to the east at naval facilities on the Delaware River; the steelworkers had been filling an order for armor plate when they walked off their jobs.

Dual private and public initiatives marked the activities of other groups. American workers in a similar fashion to businessmen organized among themselves collectively to bid up their wages and improve the general circumstances of their employment; in demanding work rules and other securities, they contributed to the bureaucratization of the society. Laboring people would also petition for state regulations on working and housing conditions; welfare policies to aid them during periods of unemployment, incapacitation, old age, and the desertion or death of breadwinners; and legal protections for trade union organizing. Farmers likewise joined together to form buying and marketing cooperatives—and they most deliberately sought government monetary policies favoring debtors, regulations on enterprises, such as the railroads, that provided them basic services, and eventually price support measures. Professionals—lawyers, doctors, and others—organized themselves as well during the late nineteenth century, founding associations that established standards for practice and limited entrance to their now hallowed ranks. Laws endorsed the activities of the new professional societies. A common denominator runs through all these stories. In the late nineteenth century, Americans of similar interest grouped themselves to undo what was perceived as ruinous competition and then utilized government further to ensure and entrench their positions. State officials also played a key role, and not only in mediating claims between groups; their careers and capacities rested on their own record of stabilizing economic circumstances. From a more overarching perspective, government officials often had to save the interested parties from themselves.

Finally, a clear story in the remaking of America in the first half of the twentieth century is found in the manifold efforts to manage the economy; concurrent attempts to create an administered politics are less evident. Elite groups were the principal actors here. At the federal level, trade associations established offices in Washington, D.C., to shape legislation directly, thus institutionalizing lobbying. The build-up of federal agencies with top-level professional staffs further divorced governmental decision-making from the popular will. At the local level, people of standing and property worked to undercut the power of neighborhood-based party bosses and competitive ward politics. They successfully pressed for structural changes in urban government; strong mayors, weakened city councils, and the centralization and bureaucratization of the provision of city services represented specific solutions for them. The two major parties also increased their administrative capacities at the time, raising huge sums of money and running campaigns through the media. The place of the local party machine in people's lives waned, and not coincidentally, voting participation at the turn of the twentieth century began a long-term decline.

To establish a more ordered politics in the face of the social unrest and political scrambles of the times, elite groups also went beyond structural reforms. They sought to eliminate portions of the electorate and deny citizenship. Immigration restrictions, regulations making voting more difficult, and most notably, the literal disenfranchisement of African-Americans and some poor whites in the South were manifestations of this course of action. The attack on immigrants, it should be noted, was not just the work of people of standing; on the West Coast, virulent opposition to Asian immigration and Asian-Americans in the late nineteenth century was spearheaded by the trade union movement. The final subduing in the West of the native tribal nations of the land represented too a political cleansing and ordering. What was created overall, then, was a new administered *political* economic regime.

How, then, does a peek into the future as of 1900 lend perspective to the nineteenth-century industrializing of America? Industrialization is part of a larger story—that of evolving political economic systems. The overturn of the mercantile regime through daily circumstance and ultimately through revolution in the late eighteenth century left the destiny of the new republic an open book. One vision for the newborn nation would have favored limited manufacture; another would have seen state-orchestrated industrial growth. Antimercantilism precluded the latter. In the former case, fear of a build-up of a propertyless proletariat continued to shape developments. However, the vision of a nation of independent, self-restrained, virtuous producers and citizens remained too weak and exclusive as an ideology to counter demographic and geographic forces that were ushering in an unfettered economic and political order; the vision also further encouraged individualist behavior. An unregulated, acquisitive market society emerged, and that spurred manufacture. Industrialization, in turn, telescoped history: with cheap products and the vast use of wage labor, industrial development quickly brought a full market society into being. Industrialization also produced economic crisis and social unrest—there was material grievance as well as unease with the corporation that had been born with the competition induced by rapid industrial growth—and the subsequent response of Americans in the late nineteenth and first half of the twentieth centuries generated a new administered political economic order. In short, industrialization was product and agent of political economic change.

Americans as they enter the twentieth-first century have expressed misgivings about the corporatist, bureaucratic, and statist world created in response to the crises of nineteenth-century industrialization. In a throwback to the Knights of Labor and the Farmers' Alliance, various protesters during the 1960s attacked large-scale private and public institutions in the name of a revived participatory democracy; conservative groups of Americans, for their part, have recently launched attacks on the growth and sway of government. These challenges to the political economic order created during the first half of the twentieth century in reaction to the upheavals of the late nineteenth have also been raised at a time when the United States is rapidly losing its industrial base and prowess. The great industrial heartland of the country built after the Civil War is now rusting and being dismantled. The American people, then, are having to deal with their history, and precisely, with the legacies of industrializing America.

Chapter 1: Context

An excellent review of research on economic and social developments in the American colonial and early republic eras can be found in John J. McCusker and Russell M. Menard, *The Economy of British America, 1607–1789* (1985). Two valuable collections of essays on economic and social life in early America are Jack Greene and J. R. Pole, eds., *Colonial British America: Essays in the New History of the Early Modern Era* (1984), and Stephen Innes, ed., *Work and Labor in Early America* (1988). A fine general text is James Henretta and Gregory Nobles, *Evolution and Revolution: American Society, 1600–1820* (1987).

The dynamics of change in New England have sparked notable debate among historians. For two recent conflicting treatments, see Christopher Clark, *The Roots of Rural Capitalism: Western Massachusetts, 1780–1860* (1990), and Winifred Barr Rothenberg, *From Market-Places to a Market Economy: The Transformation of Rural Massachusetts, 1750–1850* (1992); the debate is summarized in Allan Kulikoff, "The Transition to Capitalism in Rural America," *William and Mary Quarterly* 46 (January 1989): 121–44. Recent works by Laurel Ulrich have greatly added to our understanding of women's lives in New England in early times; these include *Good Wives: Image and Reality in the Lives of Women in Northern New England, 1650–1750* (1980) and *A Midwife's Tale: The Life of Martha Ballard, Based on Her Diary, 1785–1812* (1990). On developments in the Middle Atlantic region, see Sung Bok Kim, *Landlord and Tenant in Colonial New York: Manorial Society, 1664–1775* (1978), and James T. Lemon, *The Best Poor Man's Country: A Geographical Study of Early Southeastern Pennsylvania* (1972). A recent work that documents the intense commercial life of Philadelphia in the late eighteenth century is Thomas M. Doerflinger, *A Vigorous Spirit of Enterprise: Merchants and Economic Development in Revolutionary Philadelphia* (1986); for social life and inequalities in Philadelphia at the same time, see Billy Smith, *The "Lower Sort": Philadelphia's Laboring People, 1750–1800* (1990).

A remarkable literature has emerged on the Chesapeake region in the colonial and early republic eras. Important studies include Paul G. Clemens, *The Atlantic Economy and Colonial Maryland's Eastern Shore: From Tobacco to Grain* (1980), and Allan Kulikoff, *Tobacco and Slaves: The Development of Southern Cultures in the Chesapeake, 1680–1800* (1968). For developments in the Carolinas, see A. Roger Ekirch, *"Poor Carolina": Politics and Society in Colonial North Carolina, 1729–1776* (1981), and Daniel C. Littlefield, *Rice and Slaves: Ethnicity and the Slave Trade in Colonial South Carolina* (1981). Richard S. Dunn's *Sugar and Slaves: The Rise of the Planter Class in the English West Indies* (1972) is a seminal work on the British West Indies; another important study is B. W. Higman's *Slave Population and Economy in Jamaica, 1807–1834* (1976).

Indentured servitude as an institution is treated in David Galenson, *White Servitude in Colonial America: An Economic Analysis* (1981), and Sharon Salinger, *"To Serve Well and Faithfully": Labour and Indentured Servants in Pennsylvania, 1682–1800* (1987). Galenson also deals with the origins of slavery in the North American colonies and presents an economic argument, basically that the use of slaves lowered labor costs. A work that emphasizes the role of racial fears in the origins of slavery is Winthrop Jordan, *White over Black: American Attitudes toward the Negro, 1550–1812* (1968). A now classic study that stresses class conflict among whites is Edmund S. Morgan, *American Slavery, American Freedom: The Ordeal of Colonial Virginia* (1975).

Political economic thinking during the revolutionary and early republic periods continues to be matter of contention among historians. A key work that highlights the strength of republican or commonweal belief is Drew McCoy, *The Elusive Republic: Political Economy in Jeffersonian America* (1980). Economic liberalism and an individualist ethos is stressed in Joyce Appleby, *Capitalism and a New Social Order: The Republican Vision of the 1790s* (1984), John R. Nelson, Jr., *Liberty and Property: Political Economy and Policymaking in the New Nation, 1789–1812* (1987), and, with some qualification, in Steven Watts, *The Republic Reborn: War and the Making of Liberal America, 1790–1820* (1987). Daniel Rogers summarizes the different positions in "Republicanism: The Career of a Concept," *Journal of American History* 79 (June 1992): 11–38. The debate on manufacture that unfolded in the late eighteenth century is covered in the above books; Nelson deals with Alexander Hamilton's tangential role in boosting manufacture. Contemporary writings on manufacture by Tench Coxe, Thomas Jefferson, Alexander Hamilton, and others have been well anthologized in Michael B. Folsom and Steven D. Lubar, eds., *The Philosophy of Manufactures: Early Debates over Industrialization in the United States* (1982). Tench Coxe's career is treated in Harold Hutcheson, *Tench Coxe: A Study in American Economic Development* (1969). A history of the Society to Establish Useful Manufacture (SUM) is provided in Joseph S. Davis, *Essays in the Early History of American Corporations, Numbers I-III* (1917).

Chapter 2: Paths

The story of Samuel Slater and the creation of mill villages in Rhode Island and southeastern Massachusetts is told in Barbara Tucker, *Samuel Slater and the Origins of the American Textile Industry, 1790–1860* (1984) and Jonathan Prude, *The Coming of Industrial Order: Town and Factory Life in Rural Massachusetts, 1810–1860* (1983). The pivotal role played by Slater and other artisans in transferring technological expertise and knowledge from Britain to the United States is related in David Jeremy, *Transatlantic Industrial Revolution: The Diffusion of Textile Technologies between Britain and America, 1790–1830s* (1981). Other studies of industrialization in the countryside include Anthony F. C. Wallace, *Rockdale: The Growth of an American Village in the Early Industrial Revolution* (1978), and Judith A. McGaw, *Most Wonderful Machine: Mechanization and Social Change in Berkshire Paper Making, 1801–1885* (1987).

The building of the textile center of Lowell, Massachusetts, is treated in Robert F. Dalzell, *Enterprising Elite: The Boston Associates and the World They Made* (1987), and Thomas Dublin, *Women at Work: The Transformation of Work and Community in Lowell, Massachusetts, 1826–1860* (1979). The complicated history of shoe production in Lynn, Massachusetts, is rendered in Paul G. Faler, *Mechanics and Manufacturers in the Early Industrial Revolution: Lynn,*

Massachusetts, 1780–1860 (1981); Alan Dawley, *Class and Community: The Industrial Revolution in Lynn* (1976); and Mary Blewett, *Men, Women, and Work: Class, Gender, and Protest in the New England Shoe Industry, 1780–1910* (1988).

Philip Scranton provides a critical study of the diversified manufacturing city in *Proprietary Capitalism: The Textile Manufacture at Philadelphia, 1800–1885* (1983). The opening chapters of the following books also provide fine treatments of uneven industrial development in metropolitan areas: Susan E. Hirsch, *Roots of the American Working Class: The Industrialization of Crafts in Newark, 1800–1860* (1978), Bruce Laurie, *Working People of Philadelphia, 1800–1850* (1980), and Sean Wilentz, *Chants Democratic: New York City and the Rise of the American Working Class, 1788–1850* (1984). Information on Samuel Wetherill can be found in Miriam Hussey, *From Merchants to "Colour Men": Five Generations of Samuel Wetherill's White Lead Business* (1956); on William J. Young, in Deborah Jean Warner, "William J. Young: From Craft to Industry in a Skilled Trade," *Pennsylvania History* 52 (April 1985): 53–69; and on William Horstmann, in the William Horstmann Papers, Historical Society of Pennsylvania.

The key works on southern industrialization in the antebellum period are Fred Bateman and Thomas Weiss, *A Deplorable Scarcity: The Failure of Industrialization in the Slave Economy* (1981), and Robert Starobin, *Industrial Slavery in the Old South* (1970).

A prime example of a 1950s, Cold War–era study that attempted to draw lessons from the American past for then developing nations is Walt Whitman Rostow, *The Stages of Economic Growth: A Non-Communist Manifesto* (1960). Rostow placed great weight on the role of railroads in creating a modern economic order. Robert Fogel, in *Railroads and American Economic Growth: Essays in Econometric History* (1964), deliberately challenged this view, discounting the railroads and any notion of a "prime" mover in economic transformation. *The Economic Growth of the United States, 1790–1860* (1966), by Douglass C. North is rightfully a monumental contribution to the understanding of American economic development before the Civil War, but also represents an effort to provide developing nations of the post–World War II period with a recipe from the American past. North emphasized the importance of a key product—in the American case, cotton—as a generator of income on the world market, income that could be diffused through the society to stimulate general economic growth. North's model has been criticized on any number of grounds—he underestimated the self-sufficiency of southern plantations, he overestimated the link between northeastern and midwestern development and the income derived from cotton—but it has not been replaced by another explanation as sophisticated or systematic. A general text on the antebellum economy that takes North to task is Peter Temin, *The Jacksonian Economy* (1969).

The question of resource endowments and costs and technological innovation is treated in Nathan Rosenberg, *Technology and American Economic Growth* (1972). The key statement on the scarcity of labor thesis is H. J. Habakkuk, *American and British Technology in the Nineteenth Century: The Search for Labour-Saving Inventions* (1962). The scarcity of labor, or "Habakkuk," thesis has generated ongoing discussion and critique among economic historians; for overviews, see

Paul David, *Technical Choice, Innovation, and Economic Growth: Essays on American and British Experience in the Nineteenth Century* (1975), and Carville Earle and Ronald Hoffman, "The Foundation of the Modern Economy: Agriculture and the Costs of Labor in the United States, 1800–60," *American Historical Review* 85 (December 1980): 1055–94.

The high costs of skilled labor—particularly in fitting and assembly—is noted as a prime factor in the early shift to standardized parts production techniques in the United States. Historians recently have treated the shift as less automatic or even; continuing technical problems limited demand for mass-produced items, and labor and cultural conflict slowed and shaped developments. Important studies here are Merritt Roe Smith, *Harpers Ferry Armory and the New Technology: The Challenge of Change* (1977), and David A. Hounshell, *From the American System to Mass Production, 1800–1932: The Development of Manufacturing Technology in the United States* (1984).

Chapter 3: Reactions

The circumstances surrounding the fire at Walcott's Mill in Pawtucket, Rhode Island, is described in Gary Kulik, "Pawtucket Village and the Strike of 1824: The Origins of Class Conflict in Rhode Island," *Radical History Review* 17 (Spring 1978): 5–37. On machine-breaking in England, see a classic article, E. J. Hobsbawm, "The Machine Breakers," in E. J. Hobsbawm, *Labouring Men: Studies in the History of Labor* (1964). The generally favorable response of Americans to the new machines of the antebellum era is treated in Leo Marx, *The Machine in the Garden: Technology and the Pastoral Idea in America* (1964); John Kasson, *Civilizing the Machine: Technology and Republican Values in America, 1776–1900* (1976); Carl Siracusa, *A Mechanical People: Perceptions of the Industrial Order in Massachusetts, 1815–1880* (1979); and Bruce Sinclair, *Philadelphia's Philosopher Mechanics: A History of the Franklin Institute, 1824–1865* (1974).

Artisan life and consciousness recently has received significant historical attention. Key studies include Eric Foner, *Tom Paine and Revolutionary America* (1976); Howard Rock, *Artisans of the New Republic: The Tradesmen of New York City in the Age of Jefferson* (1979); Bruce Laurie, *Working People of Philadelphia, 1800–1850* (1980); Sean Wilentz, *Chants Democratic: New York City and the Rise of the American Working Class, 1800–1850* (1984); Charles Steffens, *The Mechanics of Baltimore: Workers and Politics in the Age of Revolution, 1763–1812* (1984); William Rorabaugh, *The Craft Apprentice: From Franklin to the Machine Age in America* (1986); Bruce Laurie, *Artisans into Workers: Labor in Nineteenth-Century America* (1989); and Ronald Schultz, *The Republic of Labor: Philadelphia Artisans and the Politics of Class, 1720–1830* (1983).

Material on labor politics in the 1820s and 1830s and the Workingmen's parties is contained in the above works, but also notably in Walter Hugins, *Jacksonian Democracy and the Working Class: A Study of the New York Workingmen's Movement* (1966), and Edward Pessen, *Most Uncommon Jacksonians: The Radical Leaders of the Early Labor Movement* (1967). The exclusive nature of artisan ideology and the contribution of organized white workingmen to racial stereo-typing in the antebellum period is well analyzed in David Roediger, *The Wages of Whiteness: Race and the Making of the American Working Class* (1991).

The protests of industrial workers in the 1830s and 1840s, particularly of women workers, is vividly rendered in Thomas Dublin, *Women at Work: The Transformation of Work and Community in Lowell, Massachusetts, 1826–1860* (1979); Mary Blewett, *Men, Women, and Work: Class, Gender, and Protest in the New England Shoe Industry, 1780–1910* (1988); and David Zonderman, *Aspirations and Anxieties: New England Workers and the Mechanized Factory System, 1815–1850* (1992).

On the notable transiency of the American people in the nineteenth century, see Stephen Thernstrom, *Poverty and Progress: Social Mobility in a Nineteenth-Century City* (1964); Stephen Thernstrom, *The Other Bostonians: Poverty and Progress in the American Metropolis, 1880–1970* (1973); and Michael B. Katz, Michael J. Doucet, and Mark Stern, *The Social Organization of Early Industrial Capitalism* (1982). Growing segmentation in American communities is treated in a number of essays in Theodore Hershberg, ed., *Philadelphia: Work, Space, Family, and Group Experience in the Nineteenth Century—Essays toward an Inter-disciplinary History of the City* (1981), and in Susan Hirsch, *Roots of the American Working Class: The Industrialization of Crafts in Newark, 1800–1860* (1978). The important influence of residential segmentation on politics is brilliantly an-alyzed in Samuel P. Hays, "The Changing Political Structure of the City in Industrial America," *Journal of Urban History* 1 (November 1974): 6–37. The emergence of competitive ward politics is treated in Amy Bridges, *A City in the Republic: Antebellum New York and the Origins of Machine Politics* (1984), and Iver Bernstein, *The New York City Draft Riots: Their Significance for American Society and Politics in the Age of the Civil War* (1990).

On wages and income disparities in antebellum America, see Robert Margo and Georgia C. Villaflor, "The Growth of Wages in Antebellum America: New Evidence," *Journal of Economic History* 42 (December 1987): 873–95, and Carole Shammas, "A New Look at Long-Term Trends in Wealth Inequality," *American Historical Review* 98 (April 1993): 412–32. On work, earnings, and family sur-vival for laboring people, see Richard Briggs Stott, *Workers in the Metropolis: Class, Ethnicity, and Youth in Antebellum New York City* (1990); Jeanne Boyd-ston, *Home and Work: Housework, Wages, and the Ideology of Labor in the Early Republic* (1990); and Walter Licht, *Getting Work: Philadelphia, 1840–1950* (1992).

Differences in cultural outlook and lifestyle among working people are treated in Bruce Laurie, *Working People of Philadelphia, 1800–1850* (1980). On working-class evangelicism and temperance reform, see Jama Lazerow, "Religion and Labor Reform in Antebellum America: The World of William Field Young," *American Quarterly* 38 (Summer 1986): 265–86, and Jed Dannenbaum, *Drink and Disorder: Temperance Reform in Cincinnati from the Washingtonian Revival to the W.C.T.U.* (1984). Evocative studies of life in the lower-class wards of the nation's cities before the Civil War are provided in Christine Stansell, *City of Women: Sex and Class in New York, 1789–1860* (1986), and Richard Briggs

Stott, *Workers in the Metropolis: Class, Ethnicity, and Youth in Antebellum New York City* (1990). Nativist attacks on immigrant and African-American workers are detailed in Michael Feldberg, *The Philadelphia Riots of 1844: A Study of Ethnic Conflict* (1975).

The classic study of the religious revivalism of the 1820s and 1830s is Whitney Cross, *Burned-Over District: The Social and Intellectual History of Enthusiastic Religion in Western New York, 1800–1850* (1950). Fervent religion figures significantly in a new and expanding literature on the emerging middle class of the antebellum era. Important works include Paul E. Johnson, *A Shopkeeper's Millennium: Society and Revivals in Rochester, New York, 1815–1837* (1978); Mary Ryan, *Cradle of the Middle Class: The Family in Oneida County, New York, 1790–1865* (1981); and Stuart Blumin, *The Emergence of the Middle Class: Social Experience in the American City, 1760–1900* (1989).

The place of women in the new middle-class family has also received significant attention from historians. The subject is treated in the above books but also notably in Nancy Cott, *The Bonds of Womanhood: "Woman's Sphere" in New England, 1780–1835* (1977). The role of women in reform movements of the antebellum period is discussed in Ellen DuBois, *Feminism and Suffrage: The Emergence of an Independent Women's Movement in America, 1848–1869* (1978), and Nancy Hewitt, *Women's Activism and Social Change: Rochester, New York, 1822–1872* (1984).

A seminal work on institution building by members of the middle class is David Rothman, *The Discovery of the Asylum: Social Order in the New Republic* (1971). Finally, there have been numerous case studies of antebellum utopian communities, particularly on the Shakers; the classic general survey is Arthur Bestor, *Backwoods Utopias: The Sectarian and Owenite Phases of Communitarian Socialism in America, 1663–1829* (1950).

Chapter 4: The Civil War and the Politics of Industrialization

The view that the Civil War represented the triumph of industrial capitalism—that the vying economic interests of the North and the South led to the war—is presented in Charles and Mary Beard, *The Rise of American Civilization*, vol. 2 (1927); Louis Hacker, *The Triumph of American Capitalism* (1940); and Barrington Moore, Jr., "The American Civil War: The Last Capitalist Revolution," in Irwin Unger, ed., *Essays on the Civil War and Reconstruction* (1968). Few historians today see the war as the result of the need or desire of northern industrial interests to control the federal government and write economic legislation on their behalf. Scholars now see the causes of the war as far more complicated, resting on political, social, and ideological tensions within and between the respective regions. Moreover, the war did not "cause" industrialization, nor is its influence on industrial expansion in the late nineteenth century clear-cut. For the war's uncertain economic impact, see David Gilchrist and W. David Lewis, eds., *Economic Change in the Civil War* (1975).

The most recent biography of Henry Clay is Robert V. Remini, *Henry Clay: Statesman for the Union* (1991); for Daniel Webster, see Robert F. Dalzell, *Daniel*

Webster and the Trial of American Nationalism, 1843–1852 (1973). The political economic position of the Whigs is treated in Eric Foner, *Free Soil, Free Labor, Free Men: The Ideology of the Republican Party before the Civil War* (1970), and Daniel Walker Howe, *The Political Culture of the American Whigs* (1979). *The Transportation Revolution, 1815–1860* (1951) by George Rogers Taylor is still the finest history of riverway improvement and turnpike, canal, and railroad building in the antebellum period. The traditional view that railroads were prime movers in the nation's drive toward sustained economic growth is presented in Leland Jenks, "Railroads as an Economic Force in American Development," *Journal of Economic History* 4 (May 1944): 1–20, and Walt Whitman Rostow, *The Stages of Economic Growth: A Non-Communist Manifesto* (1960). This interpretation is challenged by Robert Fogel, *Railroads and American Economic Growth: Essays in Econometric History* (1964); Fogel calculated the limited savings achieved in building extensive railroad rather than canal systems. For a balanced assessment, see Albert Fishlow, *American Railroads and the Transformation of the Ante-Bellum Economy* (1965). For antebellum debates on the respective merits of canals and railroads, see Julius Rubin, "Canal or Railroad? Imitation and Innovation in the Response to the Erie Canal in Philadelphia, Baltimore, and Boston," *Transactions of the American Philosophic Society* 51 (November 1961): 5–106.

The financing of mill construction in Lowell, Massachusetts, is treated in Robert F. Dalzell, *Enterprising Elite: The Boston Associates and the World They Made* (1987), and Naomi Lamoreaux, "Banks, Kinship, and Economic Development: The New England Case," *Journal of Economic History* 46 (September 1986): 647–67; family ownership and partnerships represented the norm for most industrial enterprises in the United States prior to the 1880s, as discussed in Philip Scranton, *Proprietary Capitalism: The Textile Manufacture at Philadelphia, 1800–1885* (1983). The financing of canal and railroad construction is well analyzed in Stuart Bruchey, *The Roots of American Economic Growth, 1607–1861* (1965); Harry Scheiber, *Ohio Canal Era: A Case Study of Government and the Economy, 1820–1861* (1969); Stephen Salsbury, *The State, the Investor, and the Railroad: The Boston and Albany, 1825–1867* (1967); and Arthur M. Johnson and Barry E. Supple, *Boston Capitalists and Western Railroads: A Study in the Nineteenth-Century Investment Process* (1967).

Banks and Politics in America from the Revolution to the Civil War (1957) by Bray Hammond remains the critical work on banks and economic development in the antebellum period. Also important is Peter Temin, *The Jacksonian Economy* (1969). Both books emphasize the limited role played by the First and Second National Banks of the United States and private efforts among bankers to achieve monetary stability in the young republic. The limited significance of tariffs is treated in Paul A. David, "Learning by Doing and Tariff Protection: A Reconsideration of the Case of the Ante-Bellum United States Textile Industry," *Journal of Economic History* 30 (September 1970): 521–601. On the minimal investments in public education before the Civil War, see Albert Fishlow, "Levels of Nineteenth-Century American Investment in Education," *Journal of Economic History* 26 (December 1966): 418–36.

Two classic works that emphasize the role of government—that is, state governments—in American economic development before the 1860s are Louis Hartz, *Economic Policy and Democratic Thought: Pennsylvania, 1776–1860* (1948), and Mary and Oscar Handlin, *Commonwealth: A Study of the Role of Government in the American Economy, Massachusetts, 1774–1861* (1947). Both books represent post–New Deal-era studies—that is, they point to age-old traditions of government involvement in economic affairs—yet, both also detail the eventual undoing of the commonwealth ideal and the greater privatization of economic activity. Both studies note the various state government fiscal crises that limited government initiatives. The analysis is similar to a more recent study, Harry Scheiber, *Ohio Canal Era: A Case Study of Government and the Economy, 1820–1861* (1969). An even more recent work on the subject that stresses the role of antimonopoly politics in furthering a divorce of government from the economy is Herbert Hovenkamp, *Enterprise and American Law, 1836–1937* (1991).

The early spread of the corporate form of enterprise in the United States is treated in Ronald Seavoy, *The Origins of the American Business Corporation, 1784–1855: Broadening the Concept of Public Service during Industrialization* (1982), and Pauline Meier, "The Revolutionary Origins of the American Corporation," *William and Mary Quarterly* 50 (January 1993): 51–84. Whether the Constitution of the United States is a "capitalist" document or not is discussed in a lively set of essays in Robert Goldwin and William Schambra, eds., *How Capitalist Is the Constitution?* (1982); particularly useful is an article by Forrest McDonald, "The Constitution and Hamiltonian Capitalism."

Historians recently have focused a great deal of attention on the American judiciary: judges are now seen as key government actors in the nineteenth century. Interest in the role of the judiciary and economic development was particularly sparked by the publication of *The Transformation of American Law, 1780–1860* (1977) by Morton Horwitz. Horwitz argued that local commercially minded judges, through their common law rulings, became prime facilitators of entrepreneurial activity. Horwitz's book has come under attack from many quarters, largely for its almost conspiratorial quality, but he succeeded in elevating the subject of the judiciary in historical studies.

An extremely helpful guide to legal decision making is Lawrence Friedman, *A History of American Law* (1985). An important historical work on contract law is Stanley Kutler, *Privilege and Creative Destruction: The Charles River Bridge Case* (1971); and on labor law, Christopher Tomlins, *Law, Labor, and Ideology in the Early American Republic* (1993). Herbert Hovenkamp, *Enterprise and American Law, 1836–1937* (1991), is a perceptive revision of Horwitz that notes conflict in the decision making of common pleas court judges in cases involving economic activity.

The actual economic and political economic impact of the Civil War is covered in two collections of essays: Ralph Andreano, ed., *The Economic Impact of the American Civil War* (1962), and David Gilchrist and W. David Lewis, eds., *Economic Change in the Civil War* (1965). The articles in these volumes stress the limited impact of the war on economic development. For the withdrawal of the federal government as an active agent in the economy after the war, see Morton

Keller, *Affairs of State: Public Life in Late-Nineteenth-Century America* (1977). The role of the South in continuing to block all moves toward a more centrally directed economy and polity is treated in Richard Franklin Bensel, *Yankee Leviathan: The Origins of Central State Authority in America, 1859–1877* (1990).

The view of the antebellum South as precapitalist is notably presented in Eugene D. Genovese, *The Political Economy of Slavery: Studies in the Economy and Society of the Slave South* (1965) and *The World the Slaveholders Made: Two Essays in Interpretation* (1969). An equally notable rendition of the Old South as capitalist is Robert W. Fogel and Stanley L. Engerman, *Time on the Cross: The Economics of American Negro Slavery* (1974). Recent works that emphasize republican belief in the South—that is, belief in a white man's democracy—are J. Mills Thornton III, *Politics and Power in a Slave Society: Alabama, 1800–1860* (1978), and Lacy K. Ford, Jr., *Origins of Southern Radicalism: The Carolina Upcountry, 1800–1860* (1988). For an economic analysis of the costs to the South for fighting and losing the war, see Claudia Goldin and Frank D. Lewis, "The Economic Cost of the American Civil War: Estimates and Implications," *Journal of Economic History* 35 (June 1975): 299–323. An overview of the vast literature on postbellum southern political economic developments is presented in the next section.

Chapter 5: An Industrial Heartland

Figures comparing the industrial records of Great Britain, France, Germany, and the United States in the late nineteenth century are drawn from Walt Whitman Rostow, *The World Economy: History and Prospect* (1978).

On women in the garment trades in such cities as New York and Philadelphia, see Joan Jensen and Sue Davidson, eds., *A Needle, A Bobbin, A Strike! Women Needleworkers in America* (1984). On the building of retail emporiums and large offices in metropolitan areas and the employment of women sales and office clerks, see Susan Porter Benson, *Counter Cultures: Saleswomen, Managers, and Customers in American Department Stores, 1890–1940* (1986), and Margery Davies, *Woman's Place Is at the Typewriter: Office Work and Office Workers, 1870–1930* (1982). On single women in general working in new urban centers, see Joanne J. Meyerowitz, *Women Adrift: Independent Wage Earners in Chicago, 1880–1930* (1988).

The textile centers of New England in the late nineteenth century and the immigrant workforces of the mills are treated in the following studies: Donald B. Cole, *Immigrant City: Lawrence, Massachusetts, 1845–1921* (1963); Tamara Hareven, *Family Time and Industrial Time: The Relationship between the Family and Work in a New England Industrial Community* (1981); John Cumbler, *Working-Class Community in Industrial America: Work, Leisure, and Struggle in Two Industrial Cities* (1979); and Gary Gerstle, *Working-Class Americanism: The Politics of Labor in a Textile City, 1914–1960* (1989). For Lynn, Massachusetts, see Mary Blewett, *Men, Women, and Work: Class, Gender, and Protest in the New England Shoe Industry, 1780–1910* (1988).

On the new industrial cities paralleling the Atlantic coast, see the following: Carol E. Hoffecker, *Wilmington, Delaware: Portrait of an Industrial City, 1830–1910* (1974); John Cumbler, *A Social History of Economic Decline: Business, Politics, and Workers in Trenton* (1989); Philip Scranton, ed., *Silk City: Studies on the Paterson Silk Industry, 1860–1940* (1985); Cecelia Bucki, "Dilution and Craft Tradition: Munitions Workers in Bridgeport, Connecticut, 1915–19," in Herbert G. Gutman and Donald H. Bell, eds., *The New England Working Class and the New Labor History* (1987); Jeremy Brecher et al., *Brass Valley: The Story of Working People's Lives and Struggles in an American Industrial Region* (1982); Judith E. Smith, *Family Connections: A History of Italian and Jewish Immigrant Lives in Providence, Rhode Island, 1900–1940* (1985); and Roy Rosenzweig, *Eight Hours for What We Will: Workers and Leisure in an Industrial City, 1870–1920* (1983).

Economic and social life in industrial cities along the Erie Canal in upstate New York are pictured in such works as Daniel Walkowitz, *Worker City, Company Town: Iron and Cotton-Worker Protest in Troy and Cohoes, New York, 1855–84* (1978); Brian Greenberg, *Worker and Community: Response to Industrialization in a Nineteenth-Century American City, Albany, New York, 1850–1884* (1985); and Virginia Yans-McLaughlin, *Family and Community: Italian Immigrants in Buffalo, 1880–1930* (1977). A general history of the anthracite coal region of eastern Pennsylvania is provided in Donald L. Miller and Richard E. Sharpless, *The Kingdom of Coal: Work, Enterprise, and Ethnic Communities in the Mine Fields* (1985). The importance of anthracite coal in American industrialization is convincingly stated in Alfred D. Chandler, "Anthracite Coal and the Beginnings of the Industrial Revolution in the United States," *Business History Review* 46 (Summer 1972): 141–81. Gerald Eggert, *Harrisburg Industrializes: The Coming of Factories to an American Community* (1993), offers a detailed portrait of a medium-size industrial city in south-central Pennsylvania. Key works on Pittsburgh, the steel industry, and social life and conflict in this American industrial colossus include David Brody, *Steelworkers in America: The Nonunion Era* (1960); Frank G. Couvares, *The Remaking of Pittsburgh: Class and Culture in an Industrializing City, 1877–1919* (1984); and Paul Krause, *The Battle for Homestead, 1880–1892: Politics, Culture, and Steel* (1992).

The spread of industry through Ohio is well portrayed in Raymond Boryczka and Lorin Lee Cary, *No Strength without Union: An Illustrated History of Ohio Workers, 1803–1980* (1982). Steven J. Ross has written an excellent study of industrialization in Cincinnati, *Workers on the Edge: Work, Leisure, and Politics in Industrializing Cincinnati, 1788–1890* (1985). For Detroit, see Olivier Zunz, *The Changing Face of Inequality: Urbanization, Industrial Development, and Immigrants in Detroit, 1880–1920* (1982), and Richard Oestreicher, *Solidarity and Fragmentation: Working People and Class Consciousness in Detroit, 1875–1900* (1986); for Milwaukee see Kathleen Conzen, *Immigrant Milwaukee, 1836–1860* (1976). The literature on Chicago is voluminous; William Cronon, *Nature's Metropolis: Chicago and the Great West* (1991), is a recent comprehensive work that analyzes Chicago's pivotal role in greater midwestern and western development.

Rearrangements of agricultural production in the South with the abolition of slavery and the emergence of sharecropping are well analyzed in Roger Ransom and Richard Sutch, *One Kind of Freedom: The Economic Consequences of Emancipation* (1977); Jay Mandle, *The Roots of Black Poverty* (1978); and Gerald David Jaynes, *Branches without Roots: Genesis of the Black Working Class in the American South, 1862–1882* (1986). Gavin Wright has written two books that emphasize two different causes of the South's economic misfortunes after the Civil War. In *The Political Economy of the Cotton South: Households, Markets, and Wealth in the Nineteenth Century* (1978), he stresses the devastating impact of the end of the cotton boom. In *Old South, New South: Revolutions in the Southern Economy since the Civil War* (1986), he presents an argument that rests more on the dynamics of sharecropping and the creation of an isolated low-wage economy.

An overview of industrialization in the South after the Civil War is provided in James C. Cobb, *Industrialization and Southern Society, 1877–1984* (1984). Textile mill building in the Piedmont region is described in David L. Carlton, *Mill and Town in South Carolina, 1880–1920* (1982). For developments in Birmingham, Alabama, see Carl V. Harris, *Political Power in Birmingham, 1871–1921* (1977).

The role of elites in the South after the Civil War has been hotly debated by scholars since the publication of C. Vann Woodward's *Origins of the New South, 1877–1913* (1951). Vann Woodward argued that a new commercially and industrially minded elite replaced the defeated landholding rulers of the South and ushered in a "new," modern order. Recent studies have emphasized the persistence of the old landholding class and their joining with men of commerce in overseeing limited economic development. The debate and revisions of Vann Woodward's argument are succinctly stated in Jonathan Weiner et al., "Class Structure and Economic Development in the American South, 1865–1955," *American Historical Review* 84 (October 1979): 970–1006, and Steven Hahn, "Class and State in Postemancipation Societies: Southern Planters in Comparative Perspective," *American Historical Review* 95 (February 1990): 75–98. A recent survey that attempts a balanced portrait is Edward Ayers, *The Promise of the New South: Life after Reconstruction* (1992).

On the geography of late-nineteenth-century industrialization and the importance of middle-range cities, see David R. Meyer, "Midwestern Industrialization and the American Manufacturing Belt in the Nineteenth Century," *Journal of Economic History* 49 (December 1989): 921–37. Allan Pred has provided important general studies of the geography of American industrialization. See his *Spatial Dynamics of U.S. Urban-Industrial Growth, 1800–1914* (1966) and *City-Systems in Advanced Economies: Past Growth, Present Processes, and Future Development Options* (1977). A valuable theoretical perspective is also afforded in Michael Storper, *The Capitalist Imperative: Territory, Technology, and Industrial Growth* (1989). Storper argues that the location of enterprise is not simply a function of natural locational advantages, but also the result of deliberate development schemes.

The role of immigration in late-nineteenth-century industrialization is treated in Herbert Gutman and Ira Berlin, "Class Composition and the Development of the American Working Class, 1840–1890," in Herbert Gutman, *Power and Culture: Essays on the American Working Class* (1987). The particular demographic profile of American immigrants at the time—largely males in their twenties and thirties—and the impact on economic development is discussed in Alan Taylor, "External Dependence, Demographic Burdens, and Argentine Economic Decline after the *Belle Epoque*," *Journal of Economic History* 52 (December 1992): 907–36. For the often neglected contribution of Asian-Americans to the nation's economic growth, see Sucheng Chan, *This Bittersweet Soil: The Chinese in California Agriculture, 1860–1910* (1986). Statistics on population expansion in the late nineteenth century, shifts in industrial production, and the contribution of labor force growth to gains in output are available in Lance Davis et al., *American Economic Growth: An Economist's History of the United States* (1972).

The persistence of old shop-floor practices is discussed in David Montgomery, *Workers' Control in America: Studies in the History of Work, Technology, and Labor Struggle* (1979), and Daniel Nelson, *Managers and Workers: Origins of the New Factory System in the United States, 1880–1900* (1979). The practice of inside subcontracting in factories is treated in John Buttrick, "The Inside Contract System," *Journal of Economic History* 12 (Summer 1952): 205–21. On the sway of foremen and arbitrary personnel practices even in bureaucratically managed firms, see Walter Licht, *Working for the Railroad: The Organization of Work in the Nineteenth Century* (1983). The limited impact of Frederick Winslow Taylor's experiments in scientific management is stressed in Daniel Nelson, *Frederick W. Taylor and the Rise of Scientific Management* (1980), and Daniel Nelson, ed., *A Mental Revolution: Scientific Management since Taylor* (1992).

Chapter 6: The Rise of Big Business

The great student of the American corporation in recent times has been Alfred D. Chandler, Jr. This chapter takes a more eclectic approach to the rise of the corporation than Chandler, but owes a great deal to his monumental studies. There is much information in the chapter that is drawn from his books. Chandler's most important studies include *Strategy and Structure: Chapters in the History of Industrial Enterprise* (1962); *The Visible Hand: The Managerial Revolution in American Business* (1977); and *Scale and Scope: The Dynamics of Industrial Capitalism* (1990).

Biographical information on the great business figures of the late nineteenth century can be gleaned from an old (and still provocative) classic, Matthew Josephson, *The Robber Barons: The Great American Capitalists, 1861–1901* (1934). John N. Ingham, ed., *Biographical Dictionary of American Business Leaders*, 4 vols. (1983), is also a valuable source of information. A new biography on Jay Gould is Maury Klein, *The Life and Legend of Jay Gould* (1986); for Jay Cooke, see Henrietta Larson, *Jay Cooke, Private Banker* (1936). The scheming over the Erie Railroad involving Gould, Cooke, Daniel Drew, James Fisk, and Cornelius Vanderbilt is vividly depicted in a late-nineteenth-century exposé by Charles Francis Adams, Jr., and Henry Adams, *Chapters of Erie, and Other Essays* (1886).

Andrew Carnegie has had many biographers. A recent study is Harold Livesay, *Andrew Carnegie and the Rise of Big Business* (1975). Carnegie's own writing is revealing, most notably *The Gospel of Wealth* (1886) and *The Autobiography of Andrew Carnegie* (1920). The most comprehensive study of John D. Rockefeller remains Allan Nevins, *John D. Rockefeller: The Heroic Age of American Enterprise* (1940).

Alfred Chandler's works include extensive information on Gustavus Swift and the meatpacking industry. For James Duke and the cigarette industry, see Patrick G. Porter, "Origins of the American Tobacco Company," *Business History Review* 43 (Spring 1969): 59–76.

The Morgan family, J. P. Morgan, and the Morgan banking house continue to attract the attention of writers. A recent award-winning and engaging addition to the literature is Ron Chernow, *The House of Morgan: An American Dynasty and the Rise of Modern Finance* (1990). Harold C. Passer presents a critical study of George Westinghouse, Thomas Edison, and the electrical industry in *The Electrical Manufacturers, 1875–1900: A Study in Competition, Entrepreneurship, Technical Change, and Economic Growth* (1953).

The role of appointed managers in sustaining the corporations is the major theme of Alfred Chandler's *Visible Hand*. Chandler discusses the key role of the railroads in introducing bureaucratic techniques of management in "The Railroads: Pioneers in Modern Corporate Management," *Business History Review* 39 (Spring 1965): 16–40. Diversification as a business strategy is the major theme of Chandler's *Strategy and Structure*; particularly important is his portrait of the Du Pont Company.

For the complex legal history surrounding the rise of the corporation and how antimonopoly politics spurred mergers, see Herbert Hovenkamp, *Enterprise and American Law, 1836–1937* (1991), and Tony Freyer, *Regulating Big Business: Antitrust in Great Britain and America, 1880–1990* (1992). An important study that emphasizes the role of the depression of 1893 and other contingencies in the emergence of the corporation is Naomi R. Lamoreaux, *The Great Merger Movement in American Business, 1895–1904* (1985). The critical role of finance capitalists is stressed in Gabriel Kolko, *The Triumph of Conservatism: A Reinterpretation of American History, 1900–1916* (1977). Finally, the place of labor unrest in the rise of big business is highlighted in James Livingston, "The Social Analysis of Economic History and Theory: Conjectures on Late-Nineteenth-Century American Development," *American Historical Review* 92 (February 1987): 69–95.

Chapter 7: Explosions

The history of the great railroad strikes of July 1877 is presented in Robert Bruce, *1877: Year of Violence* (1959). Vivid portraits of the most momentous labor conflicts of the late nineteenth century appear in Jeremy Brecher, *Strike!* (1977). The events surrounding the Haymarket Square bombing of May 1886 are described in Paul Avrich, *The Haymarket Tragedy* (1984). The best history of the Homestead strike of 1892 is provided in Paul Krause, *The Battle for Homestead, 1880–1892: Politics, Culture, and Steel* (1992). The community established by George Pullman is described in Stanley Buder, *Pullman: An Experiment in Industrial Order and Community Planning, 1880–1930* (1967); the standard history of the Pullman strike and boycott remains Almont Lindsey, *The Pullman Strike: The Story of a Unique Experiment and of a Great Labor Upheaval* (1942).

The most comprehensive analysis of strike statistics for the late nineteenth century is afforded in P. K. Edwards, *Strikes in the United States, 1881–1974* (1981). The historian Herbert Gutman spent his scholarly lifetime providing evidence and understanding of the community nature of labor strife. His most significant essays are anthologized in Herbert Gutman, *Work, Culture, and Society in Industrializing America: Essays in American Working-Class History* (1976), and Herbert Gutman, *Power and Culture: Essays on the American Working Class* (1987); two essays are particularly insightful on community support for workers—"Trouble on the Railroads in 1873–1874: Prelude to the 1877 Crisis?" and "The Workers' Search for Power: Labor in the Gilded Age." A study that attempts to delineate patterns of unrest by community type is Shelton Stromquist, *A Generation of Boomers: The Patterns of Labor Conflict in Nineteenth-Century America* (1987).

For developments in trade unionism in the 1860s and 1870s, see David Montgomery, *Beyond Equality: Labor and the Radical Republicans, 1862–1872* (1967). The literature on the Knights of Labor keeps growing. Recent important studies include Gregory Keeley and Bryan Palmer, *Dreaming of What Might Be: The Knights of Labor in Ontario, 1880–1900* (1982); Leon Fink, *Workingmen's Democracy: The Knights of Labor and American Politics* (1983); and Richard Oestreicher, *Solidarity and Fragmentation: Working People and Class Consciousness in Detroit, 1875–1900* (1986). Scholarly interest in Samuel Gompers and the American Federation of Labor remains equally strong. Two works that emphasize Gompers' early immersion in Marxism are William Dick, *Labor and Socialism in America: The Gompers Years* (1972), and Stuart Kaufman, *Samuel Gompers and the Origins of the American Federation of Labor, 1848–1896* (1973). Two studies that highlight the constant challenges to Gompers from within the AFL are John Laslett, *Labor and the Left: A Study of Socialist and Radical Influences in the American Labor Movement, 1881–1924* (1970), and Michael Kazin, *Barons of Labor: The San Francisco Building Trades and Union Power in the Progressive Era* (1987). Two more recent books attribute Gompers' realistic approach to trade unionism to governmental (particularly judicial) attacks on workers and their organizations; see William Forbath, *Law and the Shaping of the American Labor Movement* (1991), and Victoria Hattam, *Labor Vision and State Power: The Origins of Business Unions in the United States* (1993). The life of Gompers' counterfigure of the age, Eugene Victor Debs, is subtly rendered in Nick Salvatore, *Eugene V. Debs: Citizen and Socialist* (1982). On other radicals involved in trade unionism in the late nineteenth century, see L. Glenn Seretan, *Daniel DeLeon: The Odyssey of an American Marxist* (1979).

A long-swing approach to understanding the economic crises of the late nineteenth century is provided in David M. Gordon, Richard Edwards, and Michael Reich, *Segmented Work, Divided Workers: The Historical Transformation of Labor in the United States* (1982). An important article that points to rising real wages with no increases in productivity as the cause of falling profits and rates of growth is Jeffrey Williamson, "Late-Nineteenth-Century American Retardation: A Neoclassical Analysis," *Journal of Economic History* 33 (September 1973):

581–607; Williamson's argument is presented in book-length form in *Late-Nineteenth-Century American Development: A General Equilibrium History* (1974). For a general history of the economic fluctuations of the period, also see Rendig Fels, *American Business Cycles, 1865–1897* (1959).

Data on distribution of wealth are drawn from U.S. Department of Commerce, Bureau of the Census, *Historical Statistics of the United States, Colonial Times to 1870* (1975). On the earnings of workers in the late nineteenth century and family economics and decision making, see Peter R. Shergold, *Working-Class Life: The "American Standard" in Comparative Perspective, 1899–1913* (1982); Jeanne Boydston, *Home and Work: Housework, Wages, and the Ideology of Labor in the Early Republic* (1990); Ileen DeVault, *Sons and Daughters of Labor: Class and Clerical Work in Turn-of-the-Century Pittsburgh* (1990); and Walter Licht, *Getting Work: Philadelphia, 1840–1950* (1992). For the uncertainty and irregularity of employment in the period, see Walter Licht, *Working for the Railroad: The Organization of Work in the Nineteenth Century* (1983), and Alex Keyssar, *Out of Work: The First Century of Unemployment in Massachusetts* (1986).

The debate on the currency is treated in Irwin Unger, *The Greenback Era: A Social and Political History of American Finance, 1865–1879* (1965), and Walter K. Nugent, *Money and American Society, 1865–1880* (1968). There are vying interpretations of the farm protest movements of the late nineteenth century. A traditional work that stresses hard times as the root of protest is John D. Hicks, *The Populist Revolt: A History of the Farmers' Alliance and the People's Party* (1931). The view of farmer unrest as backward-looking and irrational is notably presented in Richard Hofstadter, *The Age of Reform from Bryan to FDR* (1955). Farm protest as a radical critique of industrial capitalism is the theme of Norman Pollack, *The Populist Response to Industrial America: Midwestern Populist Thought* (1962). In *Democratic Promise: The Populist Moment in America* (1976), Lawrence Goodwyn stresses the mobilizing and participatory aspects of the farmers' movement. A recent attempt at a synthesis is Robert C. McMath, Jr., *American Populism: A Social History, 1877–1898* (1993). For interconnections between labor and farm protest, see James R. Green, *Grass-Roots Socialism: Radical Movements in the Southwest, 1895–1943* (1978).

For the careers and writings of Henry George, Henry Demarest Lloyd, and Edward Bellamy, see John L. Thomas, *Alternative America: Henry George, Edward Bellamy, Henry Demarest Lloyd, and the Adversary Tradition* (1983). A standard work on the settlement house movement is Allen F. Davis, *Spearheads for Reform: The Social Settlements and the Progressive Movement, 1890–1914* (1967); see also Kathryn Kish Sklar, "Hull House in the 1890s: A Community of Women Reformers," *Signs* 10 (Summer 1985): 658–77. For religious and academic reformers of the age, see Charles H. Hopkins, *The Rise of the Social Gospel in American Protestantism, 1865–1915* (1967), and Dorothy Ross, *The Origins of American Social Science* (1991).

The look into the future provided at the end of the chapter represents my particular synthesis of the following works: Richard Hofstadter, *The Age of Reform from Bryan to FDR* (1955); Robert Wiebe, *The Search for Order, 1877–1920* (1967); Gabriel Kolko, *The Triumph of Conservatism: A Reinterpretation of*

American History, 1900–1916 (1977); James Weinstein, *The Corporate Ideal in the Liberal State, 1900–1918* (1968); John D. Buenker, *Urban Liberalism and Progressive Reform* (1973); Michael E. McGerr, *The Decline of Popular Politics: The American North, 1865–1928* (1986); Martin Sklar, *The Corporate Reconstruction of American Capitalism, 1890–1916: The Market, The Law, and Politics* (1988); Ellis Hawley, *The Great War and the Search for a Modern Order: A History of the American People and Their Institutions, 1917–1933* (1979); Ellis Hawley, *The New Deal and the Problem of Monopoly* (1966); Steven Fraser and Gary Gerstle, eds., *The Rise and Fall of the New Deal Order, 1930–1980* (1989); and Alan Dawley, *Struggles for Justice: Social Responsibility and the Liberal State* (1991).

Index

Library of Congress Cataloging-in-Publication Data

Licht, Walter, 1946–
Industrializing America : the nineteenth century / Walter Licht.
p.   cm.—(American moment)
Includes bibliographical references and index.
ISBN 0-8018-5013-4.—ISBN 0-8018-5014-2 (pbk.)
1. Industrialization—United States—History—19th century.
2. United States—Economic conditions—To 1865—Regional
disparities.   3. United States—Economic conditions—1865–
1918—Regional disparities.   4. Capitalism—Unites States—
History—19th century.   5. Industrial policy—United States—
History—19th century.   I. Title.   II. Series.
HC105.L53   1995
338.0973'09'034—dc20   94-37654